The Climate Crisis and Its Solution:

An Energy Transformation

Robert B. Fraser

Library of Congress Control Number: 2012903722
CreateSpace Independent Publishing Platform
North Charleston, South Carolina
ISBN-10: 1470140780
EAN-13: 9781470140786

Table of Contents

Preface

The goal of this book is to provide a brief presentation of the science of climate change and an outline of the energy technologies that will allow us to live sustainably on the earth. That is, to live in such a way that we leave the earth in a livable state for coming generations. As an added benefit, these technologies will usually lead to significant cost savings as well.

The Summary is an overall synopsis of the most important ideas and conclusions of this report. Part I outlines the science of climate change and some of the impacts we can expect to see now and in the coming decades. Part II discusses our current fossil-fuel-intensive use of energy and some possible low-carbon alternatives. Part III lists some specific steps we can take to achieve these low-carbon goals.

Appendix A lists the energy requirements and resulting emissions for a moderate-carbon and a low-carbon future and the steps that must be taken to realize the low-carbon future. Appendix B covers the transportation and electrical power sectors of the energy economy in greater detail.

To keep this report as objective as possible, each important assertion has an endnote listing a source, plus, where possible, a web reference. Additional scientific publications, government reports, books, and websites on climate change and energy technology are listed in an annotated bibliography.

The author may be contacted through his website at www. climateandenergyreport.org. A pdf version of this book can be found there as well, which makes accessing the web citations much easier.

I extend my deepest appreciation to the many friends and colleagues who have corrected errors and suggested changes that made the ideas clearer. In particular, I would like to thank Bill Parker for reading the entire text and making many helpful suggestions, both editorial and technical. Thanks to Wes Nisker, Djuna Odegard, Steven Kaplan, and Mark Sommer for helping to clarify the ideas in the summary and Part III, and to Aidan Fraser for help with the images. Finally, special thanks to Phyl Speser for hosting the original version of this report on the Foresight Science and Technology website.

The cover photo of the earth is from the archives of the National Aeronautics and Space Administration.

Summary

THE EVIDENCE

Based on temperature records, ice cores, and atmospheric physics and chemistry, the evidence is overwhelming that we are in the midst of a climate crisis brought on by the increasing accumulation of greenhouse gases (GHGs). Nearly every climate scientist is in agreement that the increase in these gases over the last 250 years is mainly the result of fossil fuel use and, to a lesser extent, altered land management. We also know that this increase in GHGs has led to an increase in the average global temperature.

THE PROBABLE IMPACTS

There are significant impacts that have occurred in the last decade and may occur in a more severe form in the coming decades unless GHG emissions are reduced:

- Historic heat waves, like the ones that hit Europe in 2003 (responsible for 30,000 deaths), Russia in 2010, Texas in 2011, and the US Midwest in 2012 — at current increases in GHG emission rates such heat waves would be considered normal in less than 40 years;
- Increased occurrences of wildfires;

- Extreme droughts, like the ones occurring in East Africa, Texas in 2011/2012, and the US Midwest in 2012;
- More extreme storms and floods, as occurred in Pakistan and Australia in 2010 and in the US in 2011 and 2012;
- Ocean acidification, leading to the loss of certain coral reefs and shell-forming creatures;
- Rising sea levels (up to 6 feet by the end of the century);
- Adverse health effects due to increases in malnutrition, intestinal diseases, and some infectious diseases;
- Mass migrations and malnutrition due to crop failures resulting from the increased range of plant diseases, drought, temperature effects, and/or changes in growing season.

A recent announcement (November 2011) by the Department of Energy notes that, in 2010, GHG emissions were up 6% from 2009, the highest annual increase seen yet. This means that the goal of keeping the average global temperature increase (relative to mid-twentieth-century levels) below 2°C (3.6°F) will be difficult to reach, and puts us on track to a 4°C-warmer world with temperatures increases over some land areas more than double the global average. Also, data collected over the last several years indicate that even a temperature increase of 2°C will lead to considerably more dangerous climate change than originally thought.

WHAT CAN BE DONE

The good news is that there are steps we can take as individuals and as governments to reduce these threats dramatically. Even if you doubt the severity, cause, or legitimacy of climate change, you will still benefit, *because the steps recommended here will make the air cleaner, will reduce our dependence on foreign oil and gas, and will make transportation, home heating, and electricity use less expensive for everyone.* Here are a few of the actions we can take (for more detail see Part III):

Individuals and Businesses

- Conserve energy—walk, bike, use public transportation.
- Drive cars with good gas mileage and eventually switch to plug-in hybrid or fully electric cars. Electric cars cost about one-seventh as much as a conventional car for in-town driving, are cheaper and easier to maintain, and produce one-sixth to one-half as many GHGs (Section III.A.3).
- Retrofit your residence or business to be as energy efficient as possible and use only energy-efficient lighting and appliances.
- Support legislation, policies, and elected officials that ensure we follow a path that reduces GHG emissions and supports low-carbon energy.

Governments

- Improve building codes to require greater energy efficiency.
- Provide tax incentives for low-carbon energy use or introduce a carbon tax that reflects the true cost of using fossil fuels; i.e., costs to the environment and public health from GHG emissions. Such tax reforms help renewable energy compete on a level playing field.
- Mandate utilities to gradually replace coal-fired and, eventually, gas-fired power plants with renewable energy (e.g., wind, solar, geothermal, and biomass—non-fossil biological material) and possibly third- or fourth-generation nuclear power plants. Improve and expand the electric grid.
- Support basic research and development in renewable energy and sustainable non-food biofuels.
- Assist low-income countries (now producing 60% of the global GHG emissions) to develop and maintain low-carbon energy technologies

REASON FOR HOPE

Rapid change is possible. Nationwide industrial mobilizations have occurred in the past and in very short periods of time. The industrial makeover during World War II is one example. Such mobilizations are happening now in the developing world, especially in China and India. This time the mobilization is toward high-tech innovations, including energy-saving and low-emission technologies that will start to reduce GHG emissions now. Even though China takes honors as the leading GHG emitter, it is also leading the world in solar and wind power, hosting the first large-scale carbon capture and sequestration site (to deal with emissions from coal-fired power plants), and helping to develop one of the first safe fourth-generation nuclear reactors.

The energy transformation that will follow from this mobilization has the potential to stabilize the atmospheric GHG concentrations to less than a 60% increase over what they were in pre-industrial times, avoiding the worst projections of climate change. As a bonus, such a transformation would create jobs, get the world off fossil fuels, and reduce the cost of energy. If we wait until the worst effects of climate change start appearing around the globe, the cost of adaptation will be many times the cost of prevention and will result in far more human suffering and ecosystem failures.

To summarize the issues: climate change due to greenhouse gas emissions is real, and its impacts are being felt now and could be considerably worse in the decades to come. The solution to this problem is a transformation in the way we use energy — i.e., toward a more efficient use of energy and a greater use of renewable forms of energy.

Part I. The Climate[1]

How do we know that greenhouse gases (GHGs) are accumulating, and how do we know that this accumulation will increase global temperatures? To establish the degree of agreement among climate experts, the publications of 1,372 climate researchers were examined and it was found that over 97% supported the tenets of human-caused climate change as outlined by the Intergovernmental Panel on Climate Change (IPCC).[2]

Part I of this book looks at GHG measurements, GHG sources, evidence of global warming, and some of the impacts expected as reported by the IPCC and other scientists.

A. Measurement of GHG Concentrations

GHGs provide a gaseous blanket that keeps the earth at a hospitable temperature of 15°C (59°F), rather than a chilly -20°C (-4°F), in much the same way that glass windows keep the air inside greenhouses warmer than the air outside.[3] The main constituent of GHG is carbon dioxide (CO_2). The earth has maintained a long-term balance between the production of CO_2 (mainly from volcanic activity before human influence) and the reduction of CO_2 (mainly from the weathering of rocks, the deposition of carbonates on the ocean floor, and the burial of organic matter to form fossil fuels).[4] The reduction in CO_2 slightly exceeded the production of CO_2 during the last 50 million years, with a corresponding decrease in global temperatures, eventually leading to the ice ages that started about 2 million years ago. During the last 800,000 years, the earth experienced a complete glacial cycle (with the ice sheets expanding then retreating) about every 100,000 years. The concentration of CO_2 in the atmosphere varied from about 180 parts per million (ppm) during glacial maximums to 280 ppm during glacial minimums (interglacial periods). The corresponding global average temperature increased 5°C to 6°C from glacial maximum to minimum. These temperature and CO_2 data were inferred from ocean sediment cores and ice cores from Greenland and Antarctica.[5]

During the past interglacial period known as the Holocene (the last 10,000 years) the concentration of CO_2 in the atmosphere had remained nearly constant at about 280 ppm until the industrial revolution about 250 years ago.[6] Since then we have been adding CO_2 at an ever-increasing rate, primarily by burning fossil fuels.

CO_2 has been monitored continuously since the mid 1950s from the mountain top of Mauna Loa in Hawaii, and from other stations around the globe. The concentration of CO_2 in the atmosphere has increased from about 317 ppm, in 1958, to 394 ppm in August of 2012.[7] This recent CO_2 measurement is higher than the highest pre-industrial concentration of CO_2 in the atmosphere during the last 800,000 years.[8] Furthermore, the *rate* at which CO_2 has been increasing in the atmosphere has accelerated from less than one ppm per year in the 1960s to about two ppm per year during the first decade of the twenty-first century.

CO_2, the most prominent GHG, is responsible for most of the increase in greenhouse effect over the last 250 years (also known as radiative forcing[9]), and remains in the atmosphere for a century or more. The atmospheric concentrations of the remaining GHGs (methane, nitrous oxide, and certain halocarbons) have all been measured, and their concentrations combined are less than 1% of the concentration of CO_2.[10] But because they are more effective than CO_2 as GHGs, their combined contribution is about 56% of the radiative forcing due to CO_2.[11]

B. Sources of GHG Emissions

Figure 1 shows the global GHG emissions from various sources due to human activities in 2010. These emissions are measured in billions of metric tons (gigatons) of carbon (GtC). This way of measuring GHG emissions only counts the weight of carbon, but not oxygen or other gases.[12] Also, the GHG emissions for gases other than CO_2 are weighted by their radiative forcing effect, so that their true potential as GHGs can be seen on the same chart as the CO_2 emissions.

The first four entries in the graph comprise CO_2 emissions from fossil fuel consumption and cement manufacturing (including a small amount due to gas flaring).[13] These emissions are of the greatest concern because of their amount and rate of increase. These emissions increased at about 1.5% per year from 1970 to 2000, then rose at over 3% per year for most of the next decade (2000 to 2007)[14] before tapering off during the recession of 2008-09. Then they jumped a record 5.9% from 2009 to 2010 (the increase in China was 10%!), and another 3.2% in 2011.[15]

The next two entries in Figure 1 comprise CO_2 emissions from land-use change. The *sources* of CO_2 emissions due to land-use change are mainly deforestation, burning, and land cultivation. The *sinks* (CO_2 drawn down into the soil) are mainly from revegetation and accumulation of organic matter in the soil. When the two are combined, there is a net source of 1.5 GtC/y, which has the remarkable property of remaining almost constant for the last fifty years.[16] This is because the increase in deforestation in the tropics is somewhat offset by the reforestation and increase in soil organic material in the temperate zone; but this may not be the case going forward, as the deforestation of tropical forests continues.[17] This threat is particularly alarming given that emission sources from land-use change already make up such a large contribution to total CO_2 emissions. How this threat can be managed is discussed in Part III.

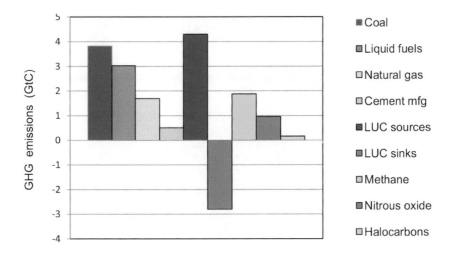

Figure 1. *Human-caused global greenhouse gas (GHG) emissions from various sources in units of billions of metric tons (gigatons) of carbon (GtC) in 2010. The first 6 entries in the chart are CO_2 emissions. The last 3 entries are emissions from GHGs other than CO_2 and have been adjusted to account for their effective radiative forcing. Liquid fuel is mainly oil and gasoline. LUC = Land-use change. See text for references.*

The methane emissions shown in Figure 1 arise primarily from landfills, livestock, rice production, and a small amount from biomass and fossil-fuel production and burning.[18] Naturally occurring methane emissions are about 40% less than this and arise from wetlands, permafrost, and methane hydrates on the ocean floor. However, as temperatures rise (particularly in the Arctic) due to human-caused global warming, the release of methane from permafrost could become a significant source of GHG emissions, since methane is about 70 times more effective than CO_2 as a GHG over a period of 20 years.[19]

The nitrous oxide emissions shown in Figure 1 arise from the use of nitrogen-based fertilizers and land and livestock management.[20] Finally, the halocarbons, which include chlorofluorocarbons (CFCs) and other halogen-containing gases, arise partly from a need to replace the ozone-depleting chlorinated gases and make up a little over 1% of all GHG emissions.[21]

CO_2 emissions (the first six entries in Figure 1) totaled 10.55 GtC in 2010 or 78% of all GHG emissions. Except for a brief discussion of proposals for reducing methane and nitrous oxide in Part III, the remainder of this report will be focused on CO_2 emissions and how they can be reduced.

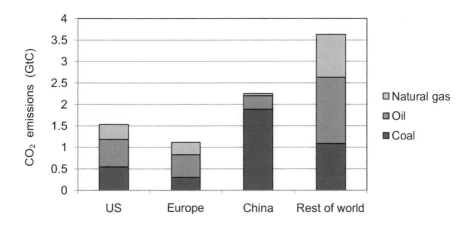

Figure 2. *Fossil-fuel CO_2 emissions for various regions of the world in units of billions of metric tons of carbon (GtC) for 2010.[22] The concern here is that China's exceptional economic growth rate will lead them to use even more coal and possibly much more oil, thus leading to dramatically higher CO_2 emissions.*

Figure 2 shows the average CO_2 emissions in 2010 for fossil fuels in various regions of the world. One important feature of Figure 2 is that China is using about half of the world's coal, which has twice the CO_2 emissions per unit of energy as natural gas. With an 8% growth rate in their GDP (2011), China would double their use coal in only 9 years. Clearly, this is unsustainable, would lead to even worse air pollution in China's main industrial centers, and would be devastating to the world's attempts to reduce global GHG emissions. Hopefully, China will develop carbon-capture-and-sequestration (CCS) technology or replace coal with natural gas, nuclear, or renewable forms of energy, perhaps with the help and encouragement of the developed world.

The second feature of note in Figure 2 is that, if China were to increase their oil use to match the rest of the world, there would be an acute shortage of oil worldwide, and global GHG emissions would be driven still higher. China can avoid this by encouraging the use of cars with very high gas mileage or electric cars, in parallel with the transformation of their electric grid to nuclear or renewable power.

Further discussion of the reduction of GHG emissions can be found in Parts II and III.

C. Evidence for the Warming Effects of GHGs

This section outlines the evidence we have that the accumulation of GHGs will increase global temperatures.[23]

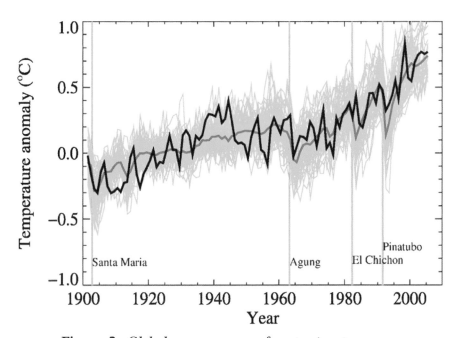

Figure 3. Global mean near-surface temperatures over the twentieth century from observations obtained from the Hadley Centre data set (black) and as obtained from 58 simulations produced by 14 different climate models driven by both natural and human-caused factors that influence climate (yellow). The mean of all these runs is also shown (red line). Temperature anomalies are shown relative to the 1901 to 1950 mean (subtract 0.15°C to get anomalies relative to the mid-twentieth century mean). Vertical gray lines indicate the timing of major volcanic eruptions.[24]

7

1. Measurements and observations

Measurements of temperature worldwide over the last 100 years have shown a clear increase of about 1.0°C (1.8°F). The steepest rise has occurred in the last 50 years, corresponding to the sharpest increases of CO_2.concentrations (see Figure 3). Recent analysis of a much larger data set of land temperature measurements going back to the eighteenth century supports the rise in temperatures over the last 50 years seen in Figure 3, [25] This recent analysis also shows a very close relationship between this temperature rise and the increasing levels of atmospheric CO_2.

Perhaps better than words or graphs is an animation, compiled by NASA*, of temperature variations throughout the world over the last 130 years. Two years from the animation are shown in Figure 4. The year 1971 is representative of the years 1950 through 1980, while the year 2010 is representative of the last few years. The red areas represent temperatures that are above the 1951-to-1980 baseline average temperature,[†] with the darkest reds being 2°C (3.6°F) or more above that baseline. The blue areas represent temperatures that are below the baseline average temperature, with the darkest blues being 2°C or more below the baseline. Note particularly the increase in temperatures in 2010 in the northern latitudes. This increase arises because temperatures in the Arctic are much more affected by global warming than are temperatures in lower latitudes, most likely because, as sea ice melts, it exposes more dark-colored, sun-absorbing sea surface in summer months. This heat is then released during the fall and early winter and can have a major impact on weather events in lower latitudes.[26]

---

* http://www.nasa.gov/topics/earth/features/2011-temps.html
† We will use this as our baseline average global temperature in this report, or, equivalently, the mid- twentieth century average global temperature.

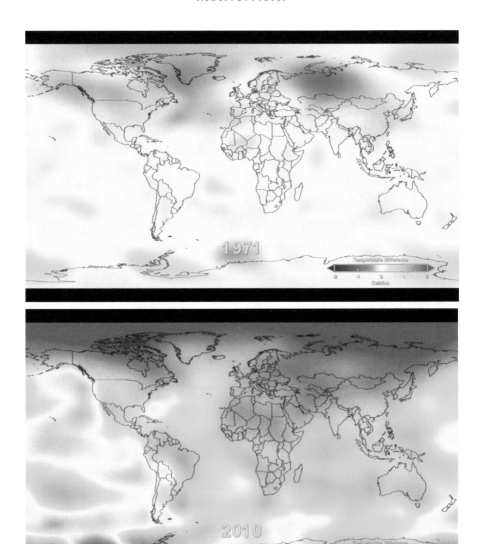

Figure 4. *Global temperature variations for 1971 and 2010 from the animation at www.nasa.gov/topics/earth/features/2011-temps.html. Reds indicate temperatures higher than the average during a baseline period of 1951-1980, while blues indicate temperatures lower than the baseline average. (Data source: NASA Goddard Institute for Space Studies. Visualization credit: NASA Goddard Space Flight Center Scientific Visualization Studio).*

Another observation that supports global warming is the number of record high-to-low temperatures. From 1800 weather stations located around the contiguous 48 states in the US, it has been observed that the ratio of the number of record highs to record lows from 1950 through 1980 was about 1:1. That is, for every record high temperature recorded in a given year, there was a record low. But that ratio increased to 2:1 during the first decade of the twenty-first century,[27] 3:1 in 2011, and 10:1 during the first half of 2012 (although 2012 will probably not continue at this pace).[28]

In addition to global temperature increase, observations that support global warming include:

- a decrease in the volume of ice in glaciers and ice sheets (leading to a rise in sea level) and a decrease in Arctic Sea ice (which accelerates the warming by exposing more dark, sunlight-absorbing ocean);
- an increase in permafrost thaw (leading to the release of the GHG methane, which increases temperatures further);
- an increase in ocean acidity due to an increase of dissolved CO_2 (which adversely affects shell-forming organisms and corals).[29]

2. Ancient climates

The temperatures and CO_2 concentrations over the last 650,000 years can be measured from samples taken at various depths in polar ice cores. From these measurements we can see how temperature varies with CO_2 concentration, including the effects of various feedback cycles in the climate system.[30] The current warming follows the same relationship to CO_2 concentrations as in this ancient record.

3. Physical models

By modeling the physics and chemistry of the climate, certain features like global mean temperature can be estimated for periods of time in the past, present, or future. (The accuracy of these predictions depends on the assumptions used, the science included, and the resolution of the grid employed in the calculations.) The confidence in these models is based[31] on their:

- reliance on known physical principles,
- ability to accurately simulate the climate of the recent past (see Figure 3), and
- ability to simulate certain features of the last glacial age that are known through ice core samples.

An example of the accuracy achieved when using models to estimate mean global temperature is shown in Figure 3. The estimated temperatures (yellow band and red line) compare favorably with the actual observed temperatures (black line) over the same time period.[32] If the GHG effects of burning fossil fuels and land-use changes are left out of the models, the rise in temperature seen in Figure 3 in the last 50 years does not appear. This adds additional support to the assertion that the recent rise in global temperature is due to the human-caused increase in GHG emissions.

Model predictions for average US (as opposed to global) temperatures this century have been carried out by the US Global Change Research Program.[33] They find that US temperature increases are somewhat higher than the global average. But more noticeable, in the higher carbon-emission scenarios, would be the frequency of heat waves (occurring up to 10 times more often than at present).

What are not factored into these model predictions are abrupt climate feedback mechanisms, such as a sudden breakup in the Greenland or Antarctic ice sheets or a sudden release of methane from the methane hydrate stores in the Arctic or Atlantic Oceans. The release of even 10% of the methane hydrates in a few years would have the effect of raising the atmospheric CO_2 level by a factor of 10.[34] Methane and

CO_2 released by the thawing of permafrost in the Arctic will add to this.[35]

Another feedback mechanism generally not included in the models is that global warming may speed up the metabolism of the most widespread marine plankton species (known as foraminifera) near the surface of the ocean. This prevents the carbon they contain from penetrating to deeper layers, thus reducing the amount of carbon that can be absorbed by the ocean. This in turn will lead to greater GHG concentrations in the atmosphere, as happened in the last hothouse period (known as the *Eocene* epoch) 34 to 56 million years ago.[36]

Other abrupt temperature changes have occurred in the past when climate tipping points were crossed. During the warming period of the last ice age, around 15,000 years ago, the Greenland temperature increased 10°C (18°F) in only three years.[37] So, in effect, the models are too conservative, since the majority of abrupt changes we are concerned about would all lead to more extreme climate conditions.

D. Impacts of GHG Emissions and Climate Change

This section will be a brief review of some of the impacts of climate change due to GHG emissions. A more comprehensive treatment can be found in the Intergovernmental Panel on Climate Change (IPCC) 2007 Working Group II document,[38] and in a more recent IPCC report emphasizing the danger of extreme weather events due to global warming.[39] The website for the Center for Climate and Energy Solutions[40] has a good section on impacts, as do two books focusing on climate change.[41,42] Also of note is that in the last several years the data suggests that even a modest increase in global temperature is much more likely to lead to "dangerous climate change" than previously thought.[43]

The following is a list of the most noticeable recent and ongoing impacts, and those predicted for later this century, if we continue our business-as-usual consumption of fossil fuels:[44]

- **More frequent heat waves** Europe experienced a heat wave in 2003 considered a "500-year" event, with 30,000 excess deaths (deaths in excess of the number expected with average temperatures).[45] A report from the US Global Change Research Program predicts that "if greenhouse gas emissions continue to increase, by the 2040s more than half of European summers will be hotter than the summer of 2003, and by the end of this century, a summer as hot as that of 2003 will be considered unusually cool."[46]

  In the summer of 2010, Russia experienced a heat wave lasting six weeks. Millions of acres of forest and thousands of homes were burned and 40% of the grain crop was wiped out.[47] As a result, world grain prices more than doubled within six months.[48] If such a heat wave was centered on the US grain production area, the effect on world prices could be far worse.[49] (A serious heat wave and drought did occur in the summer of 2012 over the US upper Midwest.) A recent study finds that there is an 80% probability that the 2010 Russian heat wave would not have occurred without global warming.[50]

  These heat waves, particularly when combined with high humidity, pose a severe public health risk, especially for the elderly. Such extreme heat events are responsible for more deaths annually than hurricanes, lightning, tornadoes, floods, and earthquakes combined.[51]

  Modern humans (*Homo sapiens*) evolved less than one million years ago, and even the genus *Homo* originated less than 3 million year ago[52], a period of time when the average global temperatures did not exceed 2°C (3.6°F) above the mid-twentieth-century average temperature. An average global temperature increase of 6°C (10.8°F) that we might see at the end of this century, if we continue to burn fossil fuels at current rates, would be far warmer than the environment in which humans evolved. Specifically, many of the heat waves could not be survived without air conditioning,[53] and crop yields in many parts of the world could not be sustained.

Paradoxically, the warming in the Arctic has caused the polar jet stream (or polar vortex) to weaken in winter, allowing large tongues of frigid arctic air to spill south, as happened in 2009-2010 in Florida and England and in the winter of 2012 in Eastern Europe and Russia. But overall the Arctic has experienced more warm areas that persist for months, as happened in Canada during the winter of 2009-2010, the warmest and driest winter since records began in 1948.[54]

- **More frequent floods**

   Heavier rains can be expected in some areas, because of the greater water-carrying capacity of the warmer atmosphere. The mid-section of the US experienced two "500-year" floods and one "1000-year" flood in the last 20 years,[55] and the Midwest and eastern US experienced an extraordinary number of tornados and flooding in the spring and fall of 2011 and the late winter/spring of 2012.

   Pakistan experienced the worst flood in 80 years in 2010, affecting 20 million people. Pakistan was hit by floods again in the early fall of 2011, and Thailand had some of its worst flooding on record in October of 2011. Australia had severe flooding in 2010-2011, due to one of the strongest La Niñas and one of the highest coastal sea surface temperatures on record.[56]

   Stronger storms can also bring higher storm surges that can cause significant damage to seaside communities, as happened along the New Jersey and New York City shorelines from Hurricane Sandy in the Fall of 2012. A 13.9-foot surge was recorded at Lower Manhattan on October 29, the highest on record.

- **Deeper droughts**

  Climate change models predict extended drought in the decades to come in some areas of Central America, the eastern Mediterranean, and Australia. The land area subjected to extreme drought is predicted to increase from about 1% today to 30% at the end of the century.[57] The southwestern US may be entering a prolonged period of drought like the one that persisted during the years 900 to 1300 A.D.[58] Severe water stress has already reached parts of the Midwest, and such conditions are expected to get considerably worse. Much of the farming in Kansas, Oklahoma, and Texas depend on the Ogallala Aquifer. In parts of this area, especially in Texas, the aquifer is dropping one to three feet per year, but its recharge rate is less than one inch per year. The continuing drought in Texas in 2011 and 2012 suggests that relief is not in sight.[59]

  In South America, the deforestation already underway in the Amazon rainforest may lead to regional climate change that in turn may transform the rainforest into savanna grassland, with an increased chance of wildfires. Global warming would significantly increase this possibility. The 2007 IPCC report warned that "the synergistic combination of both regional and global changes may severely affect the functioning of Amazonian ecosystems, resulting in large biome changes with catastrophic species disappearance."[60]

- **Shrinking Arctic sea ice and melting permafrost**

  The September 2012 Arctic sea ice extent was 49% below the 1979-to-2000 average.[61] Furthermore, the remaining ice is thinner. This ice loss will act as an amplifying effect on global warming, due to the greater exposure of darker, sunlight-absorbing ocean surface. In addition, the increased melting of the Arctic permafrost will introduce more methane and CO_2 into the atmosphere that will also increase global warming.

- **Sea level rise**

   The melting of ice sheets and glaciers and thermal expansion of the ocean will cause sea levels to rise an estimated 1 to 2 meters (3 to 6 feet) by the end of the century.[62] There were periods during the melting phase of the last glacial cycle when sea level changed by one meter (about 3 feet) in as little as 20 years. The mechanism for this much ice sheet breakup and melting is not well understood. Should this type of collapse start on the Greenland ice sheet today, it would be impossible to stop.[63] If fossil-fuel energy use continues on its current path (3% increase per year), then some countries, states, and cities may be particularly at risk for flooding by the middle or end of this century, as higher storm surges will add to the higher average sea levels. Areas at risk include The Netherlands, Bangladesh, south Florida, California's Bay Delta region, New Orleans, New York City, Venice, the coastal cities in India and China, and many others.[64]

- **Increased acidity of the oceans**

   Increased ocean acidity (resulting from increased CO_2 dissolved in the ocean) can threaten the survival of calcium-carbonate-forming shells and coral reefs. Already Australia's Great Barrier Reef is under stress and may not survive another ten years, if fossil fuel emissions continue at today's pace.[65]

- **Threatened ecosystems**

   If the global average temperature increase exceeds 3.5°C, it is predicted that there will be 40% to 70% species extinction worldwide.[66] Already it has been found that plants and animals are shifting their ranges toward the poles at almost 6 feet per day.[67] Because of physical barriers (urban areas, ranchlands, etc.), some species may not be able to move fast enough to survive. In some cases, vast expanses of forest are being lost to increasing pests and drought.[68]

- **Adverse health effects**

  The potential health impacts of climate change are summarized in the IPCC 2007 Synthesis Report. They state: "The health status of millions of people is projected to be affected through, for example, increases in malnutrition; increased deaths, diseases, and injury due to extreme weather events; increased burden of diarrheal diseases; increased frequency of cardio-respiratory diseases due to higher concentrations of ground-level ozone in urban areas related to climate change; and the altered spatial distribution of some infectious diseases."[69]

- **Mass migrations due to crop failures**

  Also noted in the IPCC Synthesis report is that "at lower latitudes, especially in seasonally dry and tropical regions, crop productivity is projected to decrease for even small local temperature increases (1 to 2°C), which would increase the risk of hunger."[70] Also of note is that plant diseases and pests can be expected to be more noticeable, while insects that provide beneficial roles, such as biological control and pollination, may not fare as well.[71]

- **Observations by intelligence agencies**

  Two reports on the impacts of climate change are from intelligence agencies, which normally one would not think of embracing environmental concerns. But they are paying attention, because they see climate change as a threat to the stability of global social structures. In the 2007 report by the Center for Strategic and International Studies (called *The Age of Consequences*), the panel that wrote the report was concerned that the global temperature rise by the end of the century (if we continue consuming fossil fuel at the current rate) would "pose almost inconceivable challenges as human society struggled to adapt. It is by far the most difficult future to visualize without straining credulity." They see the solution to this crisis as depending "on transforming the world's energy economy—America's energy economy in particular."[72] In a report by the Center for Naval Analysis (*National Security and the Threat of Climate Change*), the authors wrote: "In the national and international security environment, climate change threatens to add new hostile and stressing factors. On the simplest level, it has the potential to create sustained natural and humanitarian disasters on a scale far beyond those we see today. The consequences will likely foster political instability where societal demands exceed the capacity of governments to cope."[73]

Part II. Energy-Use Scenarios

Having established that global warming due to GHG emissions poses threats to individuals, social structures, and environmental stability, we now examine how our energy use leads to GHG emissions. Four possible energy-use scenarios are presented below. In these scenarios we will only consider CO_2 emissions, since they make up over three-fourths of GHG emissions, are easier to track than methane and nitrous oxide, and will be the target of most emission-reducing efforts. (However, techniques for reducing methane are briefly discussed in Section III.B.)

CO_2 emissions over the twenty-first century corresponding to each scenario are shown graphically in Figure 5. The atmospheric CO_2 concentrations and the best estimate for global temperature increases (relative to the mid-twentieth century) expected at the end of the twenty-first century for three of the scenarios are shown on the right side of each curve.[74] A temperature increase 50% higher than this is still considered likely, and "values substantially higher … cannot be excluded."[75] Also, temperature increases expected over land areas are about double the average global temperature increase.[76]

Some of the steps we can take to reach the goal of a low-carbon scenario are presented in Part III.

A. Business-As-Usual Scenario

In Part I of this report it was noted that CO_2 emissions increased at about 1.5% per year from 1970 to 2000.[77] Then they rose at over 3% per year for most of the next decade (2000 to 2007)[78] before tapering off during the recession of 2008-09. An announcement by the Department of Energy's Carbon Dioxide Information Analysis Center, at Oak Ridge National Laboratory,[79] notes that in 2010 fossil-fuel CO_2 emissions were up 5.9% from 2009, the highest annual increase seen yet, and another 3.2% in 2011. This seems to signal the return to increases of at least 3% per year, mainly due to the rapid growth in the economies of Brazil, India, and China.[80]

If CO_2 emissions were to increase at a business-as-usual rate of only 2.8% per year, they would triple by the year 2050 to 32 GtC/y (billions of metric tons of carbon per year) and increase to 130 GtC/y by the year 2100. Continuing this increase in the rate of fossil-fuel consumption throughout the century would be catastrophic for nearly all ecosystems on the planet and probably could not be sustained past the year 2050.

B. High-Carbon Scenario

A more realistic but still dangerous path for emissions is the one labeled "high-carbon" in Figure 5. This scenario is similar to one of the higher fossil-fuel-intensive emission scenarios described by the IPCC, which "assumes a world of very rapid economic growth [with] a global population that peaks in mid-century."[81] In this scenario, the rate of fossil fuel use increases at 2.8% per year up to the year 2050, at which point, since population has peaked, the rate of increase becomes more gradual. After 2080, fossil fuel use levels off as new low-carbon technologies kick in.[82] By the end of the century this would still result in an atmospheric CO_2 level of 1160 ppm and a best-estimate global temperature rise of 6.2°C (11.2°F) relative to mid-twentieth-century temperatures (with an upper estimate of likely temperature rise at 9.3°C).[83] A 6°C temperature rise would result in a higher global temperature than at any time in the last 15 million years[84] and would lead to some of the worst impacts listed in Section I.D – heat waves, droughts, sea level rise, crop failures, and danger of reaching climate tipping points. These adverse effects would be felt far into the future, because the CO_2 level would not return to the 2010 level (390 ppm) until after the year 3000.[85]

C. Moderate-Carbon Scenario

To avoid these dramatic and adverse effects of fossil fuel use, a concerted effort must be made to reduce the rate of increase of CO_2 emissions in the high-carbon scenario. The US Energy Information Administration (EIA) estimates global energy use up to the year 2035 for each energy source in their document: *International Energy Outlook, 2011.*[86] It estimates that, from 2010 to 2035, global average GHG emissions will increase at a rate of 1.4%/year, which is half the rate of the high-carbon scenario and which falls at around the fiftieth percentile of all emissions scenarios reported by the IPCC.[87] The projected EIA growth rate may seem low compared to the 3%/year pace that existed for most of the first decade of the twenty-first century, but the EIA may see the inevitable retraction in growth that is required by limits on natural resources.

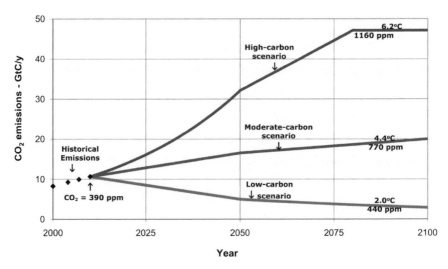

Figure 5. *Total CO₂ emissions in billions of metric tons of carbon per year (GtC/y) over a 100-year span for three carbon emission scenarios. The atmospheric CO₂ concentrations and average global temperature increases (relative to the mid-twentieth century) expected at the end of the twenty-first century are shown on the right side of each curve. For comparison, the average atmospheric CO₂ concentration for 2010 was 390 ppm.*[88]

If the global average of 3%/year increase continued for the next 40 years, the world would be consuming over three times the amount of fossil fuel that it consumes now. With coal and oil production close to peak at this time, it is hard to see how we could sustain the 3%/year growth for that long (see Section III.C for more on sustainability).

If we extrapolate the EIA 1.4% growth rate to the year 2050, then reduce the growth rate to 0.4%/year to the year 2100 (as more energy-efficient technologies kick in), the moderate-carbon scenario shown in Figure 5 (red curve) is the result. In this scenario, the world would require 920 quads (quadrillions of BTU) of energy by the year 2050, 720 of which would have to come from fossil fuels. The corresponding CO_2 emissions by the year 2050 would be 16.4 GtC/y, 14 of which are due to fossil fuel emissions, 0.9 due to cement manufacture, and 1.5 due to land-use-change.

In this moderate scenario, the estimated CO_2 level by the end of the twenty-first century would be about 770 ppm, and the global average temperature would have increased by 4.4°C (7.9°F) above mid-twentieth century temperatures (with an upper estimate of likely temperature rise at 6.6°C).[89] In a series of recent journal articles it has been concluded that the effects of a 4°C temperature rise are more severe than thought as recently as the 2007 IPCC assessments, so many of the adverse effects of the high-carbon scenario would be seen even in the moderate scenario.[90] Furthermore, a 4°C temperature rise implies a higher global temperature than at any time in the last 5 million years, when sea level was 15 to 25 meters (50 to 80 feet) higher than it is today.[91]

Therefore, even more significant cuts in emissions from fossil fuels and land-use changes must take place in the next 40 years if the global temperature increase is to be kept at or less than 2°C and the atmospheric CO_2 concentration kept at or below 450 ppm (Hansen recommends a CO_2 level below 350 ppm[92]).

D. Low-Carbon Scenarios

Energy-use scenarios that could keep the atmospheric CO_2 from exceeding 450 ppm have been described in several reports, scientific publications, and articles. We will have a brief look at a few of these scenarios, arranged by the depth of their proposed emission cuts.

1. Scenarios that would reduce CO_2 emissions by 100%

Jacobson (Stanford University) and Delucchi (University of California at Davis) recommend that 100% of the world's energy use should come from renewable sources by as early as 2030.[93] Specifically, these sources include wind, solar, hydro, tidal, wave, and geothermal power, along with existing nuclear power plants. Of course, there would be no use of fossil fuels. Also, there would be no new nuclear plants, because construction and fueling of these plants create 25 times more carbon emissions than the construction of wind turbines. No biofuels or power from biomass is allowed, because of air pollution during combustion. No carbon capture and sequestration (CCS) from coal is permitted, because of the energy needed to bury the carbon, the possible escape of CO_2 and pollution, and the problems of mining and transporting coal (see Section B.1 of Appendix B for more detail on CCS). Transportation would be powered by battery, hydrogen fuel cell, or hydrogen combustion. Heating would be by electric heat pumps and electric resistance heating. Overall, using electricity to replace fossil fuel for energy reduces the total amount of energy needed (by about one-third), because electrification is a more efficient way to use energy. But, they propose, there would still be sufficient energy for economic expansion.

The Rocky Mountain Institute proposes a similar scenario by 2050, with sustainably produced biofuels allowed and perhaps some natural gas as a transition fuel.[94]

The analysis described by Lund (Aalborg University) has Denmark going 100% renewable by 2050, using primarily wind and solar power for electricity and biomass for heating.[95]

Another approach to reducing CO_2 emissions, introduced by Pacala and Socolow (Princeton), includes the concept of "stabilization wedges".[96] They propose dividing CO_2-emitting sources into 15 different wedges on a pie chart, with each wedge representing one GtC/y of emissions. Then they propose how each source could reduce their emissions to zero. One-third of these reductions pertain to energy efficiency, one-third to alternative energy sources, including nuclear and biofuels; one-fifth to CCS from coal; and the remainder to agricultural and forestry practices.

2. Scenarios that would reduce CO_2 emissions by 80%

The British Committee on Climate Change proposes an energy use plan that allows the UK to cut its emissions by 80% below 1990 levels by 2050[97].

The California Council on Science and Technology (CCST) also proposes a scenario that would reduce CO_2 emissions by 80% below 1990 levels by 2050 (in California) as mandated by the state legislature.[98] The CCST proposes reaching their goal by aggressive efficiency measures, electrification where possible, decarbonizing the electricity supply (i.e., switching to renewable forms of electricity generation and/or nuclear power), and reducing net GHGs from the remaining fuel supply (e.g., by using sustainable biofuels).

3. Scenarios that would reduce CO_2 emissions by 50%

The goals of 80% and 100% reduction in CO_2 emissions may be too ambitious for the world to reach by 2050. A more modest goal of 50% reduction from 2010 levels will be difficult enough to achieve, but would be sufficient to keep the atmospheric CO_2 level at or below 450 ppm.

Hawkins, *et al* (National Resources Defense Council),[99] propose a scenario that limits the atmospheric CO_2 level to 450 ppm by relying in part on efficiency and renewable energy, but mostly focusing on carbon capture and sequestration (CCS) from both coal and natural gas (up to 50%). Their approach reduces the 2010 CO_2 emissions by 50% by the year 2050, then more gradually by another 20% by 2100. The problem with this approach is that it relies so heavily on the CCS technology, which has not yet been demonstrated at the large-scale electric utility level (although China and Norway have started projects) and has the ongoing problems mentioned in subsection 1 above. As a result, CCS may not be ready at the level required in this scenario by mid-century.

A low-carbon scenario without CCS

What we need is an approach that cuts emissions by 50% from 2010 levels (that is, to 5GtC/y) without relying on CCS, but which still allows a vibrant global economy. We will assume that the moderate-carbon scenario (described in Section II.C) would provide sufficient economic growth by the year 2050. But in order to transform it into a low-carbon scenario, overall energy use would have to be reduced via energy efficiencies, and most fossil fuels would have to be replaced by non-fossil energy sources.

Specifically, at least three analyses have shown that energy efficiency alone can achieve a one-third reduction in energy requirements while still allowing the same economic development.[100] So the total energy requirement for the low-carbon scenario by the year 2050 becomes 613 quads (two-thirds of the 920 quads of the moderate-carbon requirement). Most of this will have to come from non-fossil energy sources (nuclear and renewable), since CO_2 emissions in the year 2050 will have to be cut from 16.4 GtC/y (moderate-carbon) to 5 GtC/y (low-carbon). Reduction in tropical deforestation by 1.5 GtC/y by 2050 is assumed, so that the sources and sinks of land-use changes (LUC) are balanced, resulting in zero net LUC emissions by 2050.

An outline of the energy requirements and CO_2 emissions of the moderate and low-carbon scenarios, as well as a description of how to make cuts in emissions, is presented in Appendix A, and the results shown as a chart in Figures 6 and 7.

The CO_2 emissions for the low-carbon scenario are assumed to decrease linearly from 10.5 GtC/y in 2010 to 5.0 GtC/y in 2050. Emission rates are then assumed to be reduced by about 1% per year after 2050, reaching 2.9 GtC/y by 2100 (green curve in Figure 5). The 1% reduction in emissions after 2050 is consistent with the rationale for similar reductions used by the IPCC scenarios.[101]

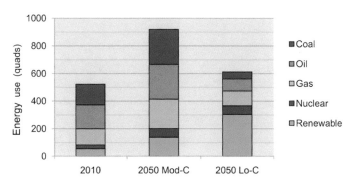

Figure 6. *Global energy consumption (in quadrillions of BTU) for 2010, for 2050 for the moderate-carbon scenario (Mod-C), and for 2050 for the low-carbon scenario (Lo-C). Note the increased use of renewable energy and reduced use of fossil fuels in the low-carbon scenario. Not shown is the high-carbon scenario which would require 1800 quads of energy per year by mid- twenty-first century and would be responsible for CO_2 emissions of over 30 GtC/y.*

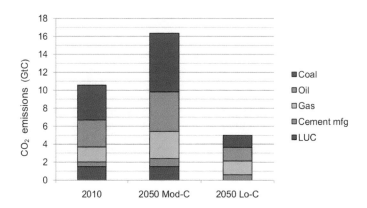

Figure 7. *Global CO_2 emissions for 2010, for 2050 for the moderate-carbon (Mod-C) scenario, and for 2050 for the low-carbon (Lo-C) scenario for the five major sources of CO_2 emissions, corresponding to the three fossil fuels shown in Figure 6 along with cement manufacturing and land-use changes (LUC – see Figure 1 and Appendix A). Units are in billions of metric tons of carbon (GtC). Note that even though the low-carbon scenario uses more energy in 2050 than in 2010, the CO_2 emissions would be down dramatically by 2050.*

As a result, the atmospheric CO_2 level in the year 2100 would rise to only 440 ppm, and the likely mean global temperature would rise by slightly less than 2.0°C (3.6°F).[102] *So the reductions of CO_2 emissions proposed by this low-carbon scenario will likely avoid the worst effects of climate change.*[103]

Admittedly, this analysis uses rough numbers, but it serves as an outline of how emissions cuts might be made in such a way as to limit the global temperature rise this century to 2.0°C, thus avoiding the catastrophic and unprecedented affects of climate change that would result with temperature gains of 4 to 6°C. (Actions that can more specifically reduce GHG emissions in the US are discussed in Part III.)

Part III. Some Actions We Can Take

We have examined the looming threat of global warming in Part I and presented some low-carbon energy strategies that could reduce this threat in Part II. Now we will look at actions that individuals, businesses, and governments can take to implement a low-carbon path. These include simple efficiency measures you can undertake for your home and business and choices you can make when buying a car or new appliance. Both the cost savings and CO_2 emission reductions of these actions are considerable.

The savings and reduction in CO_2 emissions given in this section are generally for the US only and are more detailed than the rough global estimates given in Part II.D. The US is one of the largest emitters of GHGs (second only to China), so we should be clear about how we would proceed with mitigation efforts in our own country. Furthermore, it would be in the long-term interest of the US and the world that we share emission-reduction technologies with the rapidly developing countries such as China and India. Without their participation, the struggle against unrestrained release of GHG emissions cannot be successful.

A. Actions individuals and businesses can take:

1. Choose less energy-intensive forms of transportation

Conservation and greater efficiency are often the most expedient and least expensive routes to reducing your carbon footprint (that is, the CO_2 and other GHG emissions you are responsible for). A prime example is to replace single-occupant driving with walking, biking, carpooling, or using public transportation.

2. Implement energy-saving practices in the home and business

You can make many changes to your home and workplace that will cut energy costs and reduce your carbon footprint. Many energy-saving practices are inexpensive or cost nothing at all, and yet they start saving money immediately. These are listed first, followed by actions that involve greater investment. The savings listed below are based on the average US household, including apartments; the savings could be much larger for larger homes and commercial structures. The recommendations and savings listed for each action were taken from recently published books, reports, and websites.[104]

- **Heating and air conditioning**
  Set a programmable thermostat (or manually adjust) to 21°C (70°F) for heating and 25°C (77°F) for cooling, with setbacks for nighttime or when no one is in the house. Allow sun into the windows on the south-facing side of the house in winter for passive heating. Use awnings or curtains to keep sun out during the summer.
  Savings = $100/year

- **Lighting**
  Use compact fluorescent lighting (CFL), which uses only 25% of the energy of incandescent lighting and lasts 10 times longer. Consider changing only the ten most-used light bulbs to CFLs.
  Savings = $140/year.
  Also, use natural lighting for commercial spaces when possible.

- **Water heater**
  Lower temperature to 120°F, wrap insulation around your water heater, and use a low-flow shower head.
  Savings = $80/year

- **Laundry**
  Use cold water wash, and clean the lint trap for every use.
  Savings = $60/year

- **Power strips**
  Turn off power strips (when not in use) that feed electronics with a standby feature (TVs, stereos, and computers).
  Savings = $60/year

More ambitious efforts at efficiency can be more expensive and time consuming, but may be eligible for state and federal tax credits and/or zero-interest loans from your utility. New residential and commercial buildings can be designed to use 80% less energy, and older buildings can be retrofitted to use 40% less energy.[105] Some examples include:

- **Insulation**
  Insulate walls and weather-seal doors and windows, apply low-e films to windows (available at hardware stores), and in very cold or hot environments, install double-pane windows with low-e films. This can save 30% on space heating and cooling costs.
  Savings = $250/year

- **Heat pumps**
  Heat pumps are about four times as efficient as gas heaters, and can also be used as air-conditioners. On average, a heat pump costs half as much as heating with a conventional gas heater and produces half the CO_2 emissions. (Note that the savings in using a heat pump are somewhat offset by the cost of the electricity to run it.)[106]
  Savings = $50/year

- **Energy-efficient appliances**
  For example, a new efficient refrigerator uses less than half the energy of one a decade old.
  Savings = $100/year.

- **Installation of a solar hot-water heater**
  Installing a solar hot-water heater can save two-thirds of water heating costs.
  Savings = $60/year
  Total savings = $900/year

This total is for average residences and small businesses and would be many times greater for larger homes and businesses.

3. Switch to energy-efficient cars

The average fuel efficiency for passenger cars is 22.6 mpg.[107] Here is a comparison of the annual cost of gasoline for vehicles with differing efficiencies, assuming an average annual distance of 11,450 miles and gasoline at $3.75 per gallon (spring of 2012):

- Annual cost of gasoline for the average passenger car is $1900/year
- Annual cost of gasoline for a 30-mpg vehicle is $1430/year
- Annual cost of gasoline for a 50-mpg vehicle, like a small hybrid, is $860/year

- Annual cost of electricity for a fully electric car or a plug-in hybrid running only on electricity is $275/year[108]

So we see that the cost of driving an electric car is about one-seventh the cost of driving the average conventional car.
In addition, there are far fewer moving parts in an electric car, so maintenance should be easier and less expensive (even the Li-ion batteries in the Nissan Leaf and Chevrolet Volt are guaranteed for 100,000 miles).[109] Furthermore, in the US as a whole, an electric car produces, on average, half the amount of CO_2 as a conventional car for the same distance; in the Pacific states, where electric utilities use more renewable forms of energy, an electric car produces one-sixth the amount.[110]

There were several entries to the electric car market in 2011, including the fully electric Nissan Leaf and Coda, which get around 100 miles on a full charge,[111] and the Chevrolet Volt plug-in hybrid, which can run 25 to 50 miles on its Li-ion battery and a total of 380 miles after the small internal combustion engine kicks in. For more details on batteries, see Section A.2 in Appendix B.

Note that new US federal standards will require an automaker's fleet of passenger vehicles to average 54.5 mpg by 2025.[112] It is expected that electric cars will have to be in the mix to get that kind of average mileage. The fraction of electric cars should rise in subsequent decades as they become relatively cheaper and because of their low operating cost and possible tax incentives (or a carbon tax on gasoline). If electric cars and plug-in hybrids make up half the light vehicles by 2050, and other vehicles reduce their fuel consumption by using energy efficiency measures and switching to fuel cells and biofuels, then *there could be a 50% reduction in net CO_2 emissions in the US transportation sector by the year 2050.*[113]

4. Switch to biofuels

As mentioned, some of the reduction in oil use will require greater use of biofuels and the introduction of fuel cells. The most promising source of biofuel is non-food cellulosic biomass such as switchgrass, agricultural and forestry wastes, and algae.[114] These biofuel sources are considered carbon neutral, because the CO_2 emitted during the combustion of the biomass (or its derivatives) is balanced out by the CO_2 taken up by the biomass during its growth phase. These biofuel sources are generally more efficient than corn and don't compete with food markets. The production costs look very reasonable: researchers at the US Department of Energy claim to have produced cellulosic etha-nol for only $2.25 per gallon.[115] Brazil's sugarcane ethanol is the least expensive ethanol worldwide, at $0.85 to $1.40 per gallon (but it is derived from an edible crop). Another source of bio-fuel is biodiesel from plants that can grow on marginal soils. See Section A.3 and 4 in Appendix B for more detail on fuel cells and biofuels.

5. Install solar photovoltaic panels

Federal and state tax credits and other programs have made installing solar photovoltaic panels cost effective for some homes and businesses.[116] Most homes will need a one- to two-kilowatt installation, costing about $7000 per kilowatt before tax deduc-tions; the cost is 30% less if you can claim the federal tax credit (through 2016), and even less with state credits.[117] If you are paying only 10 cents per kilowatt-hour for your electricity (the national average), then it would take 25 years to recover your cost, even after federal tax credits. This may not be financially attractive to most people. However, about 70% of the cost of solar panels lies in installation charges. A recent study found that these installation charges could be reduced to one-fifth of what they are now.[118] With these reduced installation costs, given the cost of the least expensive solar cells now on the market,[119] the cost of installing solar panels would drop to less than $2500 per kilowatt. Then the time to recover your cost is more like 8 years.

In addition to saving money, these solar installations mean a reduction in CO_2 emissions as well, since they displace the utility power that is on average over 50% due to fossil fuel burning. For additional technical details on solar panels, see Section B.8 (a) in Appendix B.

6. Support green legislation

As the signs of global warming (or "global weirding," to use *NY Times* columnist Tom Friedman's phrase) become increasingly obvious, we must engage in the public discussion of climate change and renewable energy and support public officials that enact laws and policies that reduce GHG emissions. One way to do this is to check out www.350.org, where Bill McKibben posts information about various ongoing activities and planned events that educate the public, legislators, and other policy makers. Another highly recommended website that features group action and current insights into the issues is Joe Romm's blog *Climate Progress* at thinkprogress.org. NASA has a website that lists about 40 of the major climate change websites around the world, at gcmd.nasa.gov/Resources/pointers/glob_warm.html. If you prefer to act locally, your city may have an office devoted to climate and energy issues and may offer incentives to making your home more energy efficient.

B. Actions governments and non-governmental organizations can take:

1. Mandate a carbon tax

It is argued that if the carbon in coal were taxed appropriately, the price for using it would cover the environmental and health costs due to burning it and would make carbon-free methods of producing electricity, like carbon capture, more competitive.

Specifically, a recent study found that the health-care, environmental, and economic costs of mining and transporting coal and using it to generate electricity are between one-third and one-half trillion dollars per year in the US, doubling or tripling the cost of electricity generated from coal.[120] These added costs of coal consumption are essentially an uneven tax on society, affecting the individuals and ecosystems most vulnerable to pollution.

To rectify this situation, legislators can levy a carbon tax on coal (and all fossil fuels), ranging from $55[121] to $115[122] per ton of CO_2 produced. At $115 per ton of CO_2, the tax on coal would amount to one-quarter trillion dollars per year in the US, less than the cost now realized by society. The same tax on other fossil fuels would increase the average cost of electricity by 8 to 13 cents per kWh and the cost of gasoline by about $1.00 per gallon.[123, 124] The proceeds from the tax could be returned as a payroll tax credit or a monthly dividend, a refund on utility bills, an incentive for industries to use fuels that are less carbon-intensive, and/or a subsidy for the development of renewable sources of energy.[125, 126]

There are at least some efforts underway. The Australian parliament has approved a bill that imposes a modest carbon tax of about $25 per ton. This is among the world's first national carbon-trading initiatives, along with measures in New Zealand and the European Union. As of August 2012, the effect on inflation in Australia has been less than one percent annually. The expected rise in prices for the average household is estimated to be about $10 per week, while the compensation to households will be slightly in excess of this. So, households that employ energy-efficient measures will actually be ahead economically.[127]

Europe has developed the world's first carbon trading scheme, where CO_2 emitters purchase the right to emit CO_2 from a fixed total of emissions. By 2013 emission permits will be required, and by 2020 it is expected that the permit price could reach $50 per metric ton of CO_2.[128]

There is support for a carbon tax in the US as well, from sources not altogether obvious. In the four years of making his film *Carbon Nation* (2011), Peter Byck encountered industry leaders (including energy executives), administration officials, economists, and even an editor of *The Wall Street Journal* who believed there should be a price on carbon. Although they all agree that Congress will probably not act anytime soon, the interest and momentum are there.[129]

However, California has actually taken action. They have just adopted the first cap-and-trade system in the US. It "sets limits on GHG emissions and creates market incentives to encourage oil refineries, electricity generators, and other polluters to clean up their plants."[130]

2. Switch to low-carbon electricity generation

Efforts of environmental groups

One way to delay or avoid building more coal-fired power plants is through public awareness and lobbying on the part of environmental groups. In 2007, state licenses were refused or plans abandoned for 59 coal-fired power plants in the US, and 50 more are being contested in the courts.[131] As noted by the Earth Policy Institute:

> "One of the first major coal industry setbacks came in early 2007, when environmental groups convinced Texas-based utility TXU to reduce the number of planned coal-fired power plants in Texas from 11 to 3. And now even those 3 proposed plants may be challenged. Meanwhile, the energy focus within the Texas state government is shifting to wind power. The state is planning 23,000 megawatts of new wind-generating capacity (equal to 23 coal-fired power plants)."[132]

As the price of renewable power falls and designs improve in the future, such shifts in electric power production will become politically and financially easier.

Stricter environmental laws

Other factors that encourage power generation with a reduced carbon footprint include the effects of stricter environmental laws and cheaper natural gas. Newly enacted EPA requirements that limit mercury and other emissions, along with the major drop in natural gas prices, have made it difficult for some coal-fired plants to remain operational. Some experts have estimated that 10% to 20% of the coal-fired capacity could be shut down by 2016.[133]

Another way to reduce the carbon footprint of coal-fired power plants is to capture the GHG emissions and bury them — known as *carbon capture and sequestration (CCS)*. However, no utility-level demonstration of this technology has been carried out in the US to date (see Section B.1.c in Appendix B).

State mandates for renewable energy

Certain mandates known as Required Portfolio Standards (RPS) encourage utilities to replace coal-fired and eventually gas-fired power plants with renewable energy (e.g., wind, solar, geothermal, and biomass). RPS have been adopted by 29 US states and the District of Columbia (as of May 2009),[134] and in March of 2011 the California legislature passed a bill mandating 33% renewable electric power in the state by 2020.[135]

Switching from coal to natural gas

Even as the world's CO_2 emissions continue to rise, the generation of CO_2 in the first quarter of 2012 in the US was at a 20-year low. The US Energy Information Administration cites three reasons for this reduction in emissions: a mild winter in the US, a reduced gasoline demand, and switching from coal-fired to natural-gas-fired electricity generation (natural gas [essentially methane] produces half the CO_2 that coal produces per kilowatt-hour of electricity generation).[136] The switch from coal to gas occurred because natural gas prices dropped by more than half over the last four years. This drop in price was in turn due to the dramatic increase in production from shale gas drilling in the US Northeast, Midwest, and South.

Although switching to natural gas reduces CO_2 at the power plant itself, its production may release a significant amount of methane, which is about 70 times more effective than CO_2 in trapping heat in the atmosphere (over a 20-year period). This methane release will offset some of the benefit of using natural gas during electricity production, and may actually lead to an overall increase in total GHG emissions when both production and consumption emissions are included.[137] Furthermore, this fossil fuel will eventually have to be replaced by renewable sources of energy in order to keep the global temperature increase in this century below 2°C.

Switching to nuclear energy

Nuclear power plants produce very few GHG emissions, and the newer Generation III power plants address some of the safety concerns of earlier plants. (Generation IV would be even better, but may not appear until the second half of the century). However, construction costs are relatively high, nuclear waste is no small problem, and there are significant emissions associated with the construction and decommissioning of nuclear power plants and with the mining and transport of uranium. (See endnote 172 and Section B.3 in Appendix B for more detail on nuclear power, including Generation III and IV power plants and fusion power.)

Switching to renewable energy

The most effective way to produce low-carbon electricity is to switch to renewable sources of energy. The various methods of renewable electrical energy production include hydropower, bio-power, geothermal power, wind power, and solar power. Another critical component of renewable energy is creating a "smart grid"— a project, conceived on a national scale, that would correct some of the current problems in the electrical transmission grid. A smart grid would allow utilities to better manage increased electricity demands, avoid black-outs, and facilitate the addition of distributed renewable power generation. (See Section B.10 in Appendix B).

Hydropower makes up about 8% of electricity in the US (16% globally), and it is not growing, since fewer new dams are being built (see Section B.4 in Appendix B). The remaining sources of renewable energy each make up less than 3% of total global electricity (as of 2012); however, wind and solar could each make up 20% of total electricity, even with current technology (see Sections B.7 and 8 in Appendix B for technical details and references). However, some provisions of the "smart grid" would probably have to be implemented to accommodate this level of renewables.

Although biopower is a small part of electricity production today, it is expected to provide 10% of electricity in the US by 2050.[138] Finally, the US Geological Survey estimates that there are between 3.7 and 16.5 GW of potential conventional geothermal power, which would be about 3% to 15% of US electrical generating capacity.[139] But hundreds of times more than this could be generated with enhanced geothermal (although this technology has not yet been developed).[140] (See Sections B.4 through 10 in Appendix B for a more detailed discussion of renewables).

If the US can meet its goals for wind and solar (20% each), and biopower (10%); maintain the same fraction of nuclear power and hydropower as it does today; increase the share of geothermal to 7%; and reduce coal-derived electricity by 85%, then the distribution of electrical energy sources for the US in 2050 would look like:

Wind	*20%*
Solar	*20%*
Biomass	*10%*

Hydro	*7%*
Geothermal	*7%*
Nuclear	*20%*
Coal	*7%*
Natural gas	*9%*

As a result, 84% of electrical energy generation would be carbon-free by 2050 while still meeting the country's energy needs, whereas today only 30% is carbon-free (although several states in the US West have a higher percentage of carbon-free electricity generation).

3. Reduce GHG emissions from land use and waste management

Forest management

As mentioned in Section I.B, land-use changes are a significant source of CO_2 emissions. Actually, these emissions are the combination of sources and sinks. The *sources* are mainly from deforestation and land cultivation. The *sinks* are mainly from revegetation and added organic matter in the soil.[141] When these sources and sinks are combined worldwide, there is a net source of CO_2 emissions (about one-sixth of all global CO_2 emissions), which has the remarkable property of remaining almost constant for the last 50 years. This is because the increase in deforestation in the tropics (which increases CO_2) has been offset by the reforestation and increase in soil organic material in the temperate zone (which reduces CO_2).[142] However, in recent years there has been a dramatic rise in tropical deforestation that may outstrip sinks in the temperate zone. As a result there are efforts worldwide to provide incentives to slow this deforestation. One such program is REDD — Reducing Emissions from Deforestation and Degradation. Two countries that have had success in this area are China and India.[143] Other groups that are working to reduce tropical deforestation include The Nature Conservancy (www.nature.org), Rainforest Action Network (ran.org), Mongabay (rainforests.mongabay.com), and Rainforest Rescue (www.rain-forest-rescue.org).

Agricultural practices

No-till farming, in which the soil is left undisturbed and crop residue is allowed to accumulate, is another practice that reduces CO_2 emissions. By this means the nutrient and carbon content is increased at all depths relative to conventional farming, and thus it helps to reduce greenhouse gases.[144] In addition, this approach to farming increases soil aggregation and water-holding capacity and reduces soil erosion.[145] The practice has increased dramatically in the US, Canada, Brazil, Argentina, and Australia.[146] A small sample of the agencies and groups that are focused on this include the US Department of Agriculture[147], Hiwassee River Watershed Coalition No-Tillage Farming (www.hrwc.net/notill.htm), and Soil Quality for Environmental Health (soilquality.org).

Reduction of methane emissions

Methane is the most abundant GHG after CO_2. It contributes about 28% as much to the increase in the greenhouse effect (radiative forcing) as does CO_2 and (as mentioned in Section III.B.2 above) is about 70 times more effective than CO_2 in trapping heat in the atmosphere over a 20-year period. Methane is released from gas pipeline leaks, landfills, and waste treatment facilities among other sources. In a recent study it was shown that methane may have a more damaging effect than previously thought, and it may be cheaper to cut methane emissions than to cut CO_2 emissions. In fact, a 40% reduction in methane emissions is feasible at relatively low cost.[148] A variety of methods have been devised for trapping and utilizing methane.[149]

Reduction of nitrous oxide and other emissions

Although nitrous oxide is present in very small quantities in GHG emissions, it is 300 times more effective than CO_2 in trapping heat in the atmosphere. Its major sources are nitrogen-based fertilizers and soil and livestock management. There are ongoing projects studying the effective use of fertilizer and measuring the effect of soil management on the production of nitrous oxide.[150]

The halocarbon emissions can be reduced by over 50% by reducing leakage in air conditioners and refrigeration units and by reducing the use of those gases in new equipment.[151]

4. Support basic research and development in renewable energy technology

Research and development in renewable energy and sustainable technologies is being supported at the international, national, and regional level. A few examples include:

- The United Nations Department of Economic and Social Affairs (DESA), which works with governments on their renewable energy strategies, is an example at the international level. Some of their efforts include helping to attract investment in renewable energy technologies, advocating for their commercialization, and finding site-specific practical applications.[152]
- German federal energy policies have been credited with expanding Germany's installed renewable energy capacities in wind, solar, and biomass by remarkable multiples. For example, solar photovoltaic installations have increased by over 15,000% from 1990 to 2005. There are concerns, however, that the early and rapid expansion has come at the expense of basic research spending that would ensure the ongoing development of renewable technologies.[153]

- The US Department of Energy supports research and development directly through the grants and contracts program, but they also have subsidiary agencies that support renewable energy technology. For example, the National Renewable Energy Laboratory is "dedicated to the research, development, commercialization, and deployment of renewable energy and energy efficiency technologies." Another is the Office of Energy Efficiency and Renewable Energy (EERE), which "invests in clean energy technologies that strengthen the economy, protect the environment, and reduce dependence on foreign oil."[154]

5. Transfer low-carbon technology to low-income countries

The less-developed countries, most notably China and India, produce 60% of the GHG emissions, and this fraction has been increasing in recent years. Therefore, there must be international efforts to assist these countries in getting the necessary low-carbon energy technologies needed to reduce their emissions. The US has been supporting technology transfer programs like this with funding of over $1 billion made possible by the Consolidated Appropriations Act of 2010. International organizations that also support climate-related programs for low-income countries include the UN Environment Program and the Organization for Economic Cooperation and Development, among others. [155]

C. Toward a sustainable society

The central message of this report is that reducing our carbon footprint is essential to avoiding the worst effects of climate change. But there is the larger issue of creating a sustainable global society; that is, a society that meets the needs of the present generation, lives within the carrying capacity of supporting ecosystems, and can be sustained over future generations without exhausting the resources of the planet.[156] In order to maintain such a society, we must stabilize world population, reduce poverty, and protect our ecosystems, *in addition* to reducing our carbon footprint.[157]

This subject is too vast to be covered here in any detail, but a few words may be helpful. According to the Global Footprint Network, "humanity uses the equivalent of 1.5 planets to provide the resources we use and to absorb our waste." We are now on track to use 2 earth planets by 2035.[158] For example, China has enjoyed a 10%/year average growth rate over the last two decades,[159] and it now uses half of the world's coal output.[160] Even with their slower growth of "only" 8%/year in 2011, they would double their demand for coal in only nine years, leaving little for other fast-growing countries like India. Clearly this is not sustainable, since the world may be reaching its peak in coal production now.[161]

Other activities that are not now sustainable include certain kinds of ocean fishing, deforestation, mining of certain critical resources, overuse of fresh water in many areas, fossil fuel consumption, and population growth in some regions.

On a regional scale, history provides many examples of societies overusing their ecosystems to the point of collapse; e.g., the Vikings in Greenland, the Anasazi in the US Southwest, and the Mayans in Central America.[162] Because of globalization, economic interdependency, and finite mineral and biological resources, our social systems are susceptible to collapse just as these historical ones were. Indeed, we have pushed beyond what is sustainable even now.

On the other hand, we have the knowledge of history, a diffusion of environmental knowledge, access to innovative technology, and the ability for massive mobilization (like was seen in the US during World War II) that these earlier societies did not have. We have the ability to create and maintain a sustainable society. We only require the vision and will to do so.[163-168]

Appendix A

Energy Requirements and CO_2 Emissions for the Moderate and Low-Carbon Scenarios for the year 2050.*

Source of energy or emissions	Mod-C scenario		Low-C scenario	
	Energy (quads)	Emissions (GtC)	Energy (quads)	Emissions (GtC)
Coal[169] Primarily used in electricity generation. Can be cut by 80% by using renewable replacements in the form of wind, solar, hydro, geothermal, and biopower. No CCS technology required.	253	6.55	51	1.32

---

* This table gives the global energy requirements in quads (quadrillions of BTU) and CO_2 emissions in GtC (billions of metric tons of carbon) for the year 2050 for the moderate-carbon (Mod-C) and low-carbon (Low-C) scenarios. To change the units of measurement for emissions to $GtCO_2$, multiply GtC by 3.67 (see Endnote 12). The left-most column describes the cuts that must be made to the Mod-C scenario in order to realize the Low-C scenario. The energy requirements for the Mod-C scenario are obtained by extrapolating the EIA global energy estimates (Endnote 86) to the year 2050. See Section II for further description of scenarios. See Figures 6 and 7 for graphical representation of this table.

Source of energy or emissions	Mod-C scenario		Low-C scenario	
	Energy (quads)	Emissions (GtC)	Energy (quads)	Emissions (GtC)
Oil[170] Can be cut by two-thirds by requiring that: • Fully electric cars and plug-in hybrids make up half of the cars and light trucks, • Larger trucks reduce their emissions by 50% by improved efficiency and/or by the use of fuel cells or sustainable biofuels , • The residential, commercial, and industrial sectors switch from oil to electricity or improve efficiency of oil use.	253	4.40	84	1.46
Natural gas[171] Can be cut by half by implementing revised building codes for new buildings and retrofits and by the introduction of heat pumps and solar hot water heaters.	213	3.02	107	1.51
Nuclear energy[172] No change from the moderate-carbon level.	63	None	63	None

Source of energy or emissions	Mod-C scenario		Low-C scenario	
	Energy (quads)	Emissions (GtC)	Energy (quads)	Emissions (GtC)
Renewable electricity[173] Must make up for the reduction of coal use and the increased need for electrification of the economy; e.g., increased charging demands from electric cars. For the low-carbon scenario, renewable requirements are adjusted to maintain the total global energy demand for 2050 at 613 quads.	138	None	308	None
Emissions from cement manufacturing[174] Can be cut by 20%.		0.89		0.71
Net emissions from land-use change and soil management Can be reduced to zero by balancing deforestation with reforestation and improvement of soil management practices.		1.50		None
Total energy requirements and CO$_2$ emissions for the year 2050	920	16.36	613	5.00

Appendix B

Transportation and Electrical Energy Production

Transportation and electricity generation are the two sectors of the US economy that have the highest CO_2 emissions (31% and 40%, respectively, of total CO_2 emissions).

A. Transportation

The transportation sector accounts for 31% of the total CO_2 emissions due to fossil fuel combustion in the US and nearly all of these emissions are from internal combustion engines using oil/gasoline. In the US, the cost of imported oil plus the hidden costs (e.g., cost of Persian Gulf defense alliances and loss of domestic investment) accounts for over 800 billion dollars every year. *If this same amount were used in the US economy instead, it could offset the combined cost of developing renewable energy resources, a smart grid, high-speed trains, and electric vehicles.*[175]

The following are some methods for reducing both our reliance on oil and the resulting emissions. Generally, these methods also result in considerable cost savings to the individual.

1. Fuel efficiency

Fuel-efficient cars can get 50 mpg or better, compared to 22.6 mpg for the average passenger car.[176] A 50-mpg car would save over $1000 per year (at 11,450 mi/year and $3.75/gal for gas), and the US would reduce its total energy consumption by about 10%.[177] An example of such a vehicle is the electric-hybrid Toyota Prius, which gets 50 mpg to 60 mpg.

2. Electric and plug-in hybrid-electric vehicles

The most dramatic reduction in oil consumption will come from replacing conventional cars and light trucks (which create 60% of the transportation-derived CO_2 emissions) with electric or plug-in hybrid-electric vehicles.

Besides lower operating costs (see Section III.A.3), there are far fewer moving parts in an electric car, so maintenance should be easier and less expensive (even the Li-ion batteries in the Nissan Leaf and Chevrolet Volt are guaranteed for 100,000 miles).[178] Electric cars have a simpler motor and transmission design than cars with an internal combustion engine. For example, the Tesla model S has a 3-pole induction motor that has no rare-earth elements (and, hence, no material-acquisition problems),[179] weighs only 70 lbs, and requires only two speeds forward for the transmissions (one reverse) and no clutch.[180] The electric car has no heavy engine block, complicated fuel-air mixing system, emissions control, or exhaust system.

The performance of plug-in hybrids and all electric vehicles will improve as the batteries improve. Greater energy density (energy per kg of battery weight) is expected from a thin-film printing method of the Li-ion battery, a method currently under development by Planar Energy, a company spun out of the National Renewable Energy Laboratory in 2007.[181] They expect the battery to have three times the energy density of conventional Li-ion batteries at one-third the cost.

Another promising technology is the Li-air battery being developed by, among others, Risø DTU in Denmark. It has the potential of reaching the energy density of the internal combustion engine, or about four to five times the energy density of conventional Li-ion batteries.[182] This would allow a vehicle to travel 500 miles on a single charge. Strategically placed charging stations then would allow long-distance travel with no fuel-based impediments.

The ultra-capacitor is a form of energy storage that relies on charge accumulation on sheets of foil separated by microscopic distances. They are used now for storing energy for specialized needs such as regenerative braking. However, research is underway to use them to replace the batteries in electric cars. They can be discharged quickly to provide acceleration and can be charged faster than batteries.[183]

3. Fuel cells

Fuel cell vehicles offer a dramatic reduction of CO_2 emissions if the fuel is produced by renewable means. Essentially, the energy is stored as a gas, liquid, or solid and then released as electrical energy in the fuel cell. An example of an appropriate fuel would be hydrogen produced by renewable energy such as wind. In the fuel cell the hydrogen is catalytically split into hydrogen ions (protons) and electrons, to provide electrical current like a battery. It produces no tailpipe CO_2 emissions or pollutants, only water. Another appropriate fuel would be a liquid fuel produced renewably, such as methanol or ethanol grown from a sustainably produced biomass like algae. Several challenges must be met before fuel-cell cars can be competitive, including fuel production, distribution and storage; fuel-cell technology; and vehicle engineering.[184]

4. Biofuels

To eliminate CO_2 emissions due to heavy transportation (trucks, buses, air travel, and water travel), the use of biofuels (and possibly fuel cells) will need to increase. Biofuels refer mostly to ethanol and biodiesel and are usually used to replace some (10% to 85%) and, less often, all of the gasoline and conventional diesel used in IC engines.[185] Biofuels are considered carbon neutral in that the CO_2 emitted during combustion is balanced by the CO_2 the biomass consumes during its growth phase. However, to assess the actual carbon footprint of a biomass source, the entire chain of biofuel production must be considered, including fertilizer, transport of biomass, emissions during production and distribution, fuel station emissions, and combustion emissions.

As an example, switchgrass creates 4 units of energy for every unit of energy put into its production, whereas corn creates only 1.28 units of energy for each unit input.[186] [187] In fact, it is questionable if corn-based ethanol actually reduces GHG emissions at all. Another problem, in the case of corn-based ethanol, is that converting so much corn into ethanol raises the price of corn worldwide to such an extent that it presents a real hardship on poorer countries and populations that purchase corn and corn-based foods.

Some methods of producing biofuels that look more promising than corn include: cellulosic biomass (such as switchgrass and Miscanthus), agricultural and forestry wastes, algae, and Jatropha (which grows on marginal soils and is converted to biodiesel). Producing ethanol from such materials requires additional processing to break down the cellulosic materials into sugars. Considerable research is now underway worldwide to produce these biofuels at competitive prices. Researchers at the Department of Energy claim to have produced cellulosic ethanol for only $2.25 per gallon (equivalent to $120/barrel oil).[188]

The process for making ethanol from sugars or starches, on the other hand, has been known for decades. Brazil's sugarcane ethanol is the least expensive ethanol worldwide: $0.85 to $1.40 per gallon, when all expenses are considered, 30% less than US corn-based ethanol, and comparable to oil at $40 to $50 per barrel.[189]

The IEA has estimated that "the global share of biofuel in total transport fuel would grow from 2% today to 27% in 2050," and in a sustainable way that does not compromise food security.[190]

5. Alternative transportation

Finally, the least costly method of avoiding transport GHG emissions is by walking, riding a bicycle, carpooling, or taking public transportation.

B. Electrical Energy Production

Electricity generation is responsible for about 40% of the total CO_2 emissions in the US. Figure 8 shows how fossil fuel is the dominant fuel source for electricity production. Furthermore, in the US (as of March 2012) coal alone is responsible for 34% of electricity production and 67% of the CO_2 emissions caused by electricity production. In China nearly all the fossil fuel used for electricity production is coal.

Therefore, to reduce CO_2 emissions significantly, either coal must be replaced as a fuel source with renewable or nuclear energy or the CO_2 that coal produces must be buried. Currently, hydropower makes up the majority of renewable energy output (65% in the US and 85% globally), but further development is limited. So non-hydro renewable electricity and possibly nuclear power will have to expand considerably to replace coal. In 2009 renewable electricity (including hydro) constituted only 10.7% of total US electricity generation (12.7% in 2011) and 19.1% of world generation (Figure 8). This can be increased considerably, even with today's technology; Germany generates 17% of its electricity from renewables (with a goal of 35% by 2020)[191] and California generates 35% from renewables (including large hydropower).[192]

In addition, the costs for installing renewable power systems are becoming more competitive. For example, the cost of installing new wind power plants is comparable to the cost of installing coal-fired power plants, even without tax incentives.[193]

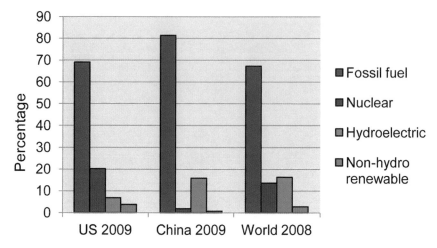

Figure 8. Electricity generation from each energy source as a percentage of the total generation in each region.[194] Fossil fuel sources include coal, natural gas, and oil.

Figure 9 shows a breakdown of non-hydro renewable electricity generation. Wind power is the major non-hydro renewable source in the US, at 46%. In China, which in 2010 passed the US in wind power capacity, wind power makes up almost 90% of non-hydro renewable electricity.

In the US, biomass energy (including landfill, agricultural and forest waste, and [rarely] municipal waste) is a close second, at 43% of non-hydro renewables. Although solar constitutes only about 0.5% of non-hydro renewables in the US, it is expected to grow by a factor of 20 to 40 by 2020, even exceeding wind power capacity.[195] China is the leading producer of solar photovoltaic cells and is committed to having solar and wind as a significant part of its electrical power. It is possible that wind and solar energy could each make up 20% of total electrical energy output in the US, Europe, and possibly China by 2030.[196] Because of the intermittent nature of wind and sun, though, meeting this goal requires the development of an updated electrical distribution grid and possibly some form of storage.

Renewable energy (especially solar) also has a place in providing water via desalination in drought-stricken areas.[197] Water is likely to become a very valuable commodity in such areas in the decades ahead. China is deploying some of the most advanced desalination plants in great numbers along its east coast.[198]

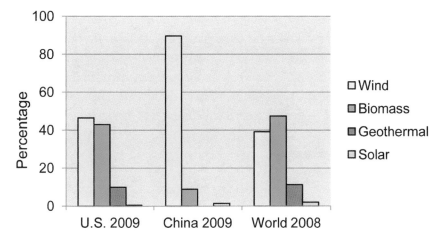

Figure 9. Electricity generation from various non-hydro renewable sources as a percentage of the total non-hydro renewable electricity for that region. Biomass consists primarily of landfill, agricultural and forest waste, and (rarely) municipal waste.[199]

In the following sections, various methods for generating electricity are considered with emphasis on the carbon footprint, cost of energy, maturity of the technology, and intermittency and storage issues.

1. Coal

As recently as 2009 coal-produced electricity constituted nearly 50% of the total electricity generated in the US. However, by March of 2012 that share had dropped to 34% and the share of natural-gas electricity had risen to 30%, as a result of a drop in gas prices. Nevertheless, coal still produces about two-thirds of the CO_2 emissions from electric power production in the US and 75% of the worldwide emissions.[200]

There are at least three ways to manage this difficulty (the first two were covered in Section III.B):

(a) Replace coal-fired power plants with renewables
Coal has the highest GHG emissions per unit of energy of any fossil fuel. Since coal is burned mainly in electricity-generating power plants, they should be the first targets for change (see Section III.B.2).

(b) Carbon tax
This is the most equitable way to get us off our addiction to fossil fuels (see Section III.B.1).

(c) Carbon capture and sequestration (CCS)
Another method being considered for reducing CO_2 emissions from coal is to capture and bury the CO_2 it generates (carbon capture and sequestration—CCS). In an MIT study of coal, the authors recommended a carbon tax as an incentive to CCS. The tax would rise from $25 in 2015 at 4% per year to $100 per ton of CO_2 by 2050.[201] By 2050 they predict that, with this tax, 60% of coal would be subject to CCS.

A technique for CCS discussed by the geologist Wallace Broecker[202] involves binding CO_2 to rock, forming permanent carbonates without the danger of leakage that one gets with CO_2 in gaseous form. There are more than enough suitable rocks—known as *ultramafics*—to bury all the CO_2 we produce. A pilot project is underway in Iceland.

Unfortunately, no utility-scale CCS facility is currently operating in the US, although some are planned. Geologists and other experts in CCS are working with the Chinese to develop large demonstration sites,[203] and Norway has built a large experimental facility at the Mongstad oil refinery.[204]

In the meantime laws requiring CCS and/or mandating a carbon tax or a cap-and-trade system are being considered. Norway has imposed a tax on CO_2 emissions for the last 20 years, which has resulted in two off-shore CCS facilities.[205] Additional details of CCS technology and policy can be found in the 2007 IPCC[206] report and in Pittock[207] and Alley.[208]

2. Natural gas

As its relative cost has gone down, natural gas has more than doubled its share of the total fuel supply for electricity generation in the last 20 years.[209] It constitutes about 23% of both US and world energy consumption.[210] Natural gas is an attractive bridge from coal to nuclear/renewable, because it has about half the carbon emissions per energy unit that coal has.[211] Fuel switching can be an effective way of changing coal-fired power plants to gas-fired.

There has been a dramatic increase in gas production in the US from shale gas; however, there are potential drawbacks as discussed in Section III.B.2.

3. Nuclear power

There has been a renaissance of interest in nuclear power in recent years because it provides base-load power (not intermittent, like wind) without GHG emissions. But this enthusiasm may be somewhat dampened by the recent nuclear power plant disaster at Fukushima, Japan. Germany has decided to close down their nuclear reactors in response to public opinion. They plan to base their energy on renewables like wind, solar, and biomass, with increased emphasis on improving their aging and under-performing electric grid.[212] *The Economist* reviewed the international state of nuclear power recently, in view of the Fukushima disaster, comparing nuclear to other low-carbon electric power sources like wind and solar. Without minimizing its challenges, the magazine did see a place for nuclear power in a low-carbon energy strategy, but growth would be very limited.[213]

In addition to the safety risks, there are problems with the high cost of construction, the disposal of nuclear waste, and the significant emissions associated with the construction and decommissioning of the nuclear plant and with the mining and transport of uranium.[214]

To put the safety of nuclear power in perspective, the International Energy Agency has collated several studies showing that nuclear power accounted for 120 deaths per trillion kilowatt-hours (TkWh) of power generated (including Chernobyl), whereas coal power accounted for 3300 deaths per TkWh[215] (the US uses about 4 TkWh of electrical energy per year). In addition, as mentioned in the section above on coal, the health-care, environmental, and economic costs of mining, transporting, and burning coal for electricity are between one-third and one-half trillion dollars per year in the US, or $1000 per person.[216]

At present, all of the reactors in the US are of an early design (Generation II) and supply about 21% of electric power in the US (27% globally). The new Generation III plants that are proposed for current construction have simpler designs, reduced capital cost, are more fuel efficient, and are inherently safer and less-complicated to operate.[217] Designs of this type are being built in China and are in the early phases of construction in the US.

Generation IV nuclear power plants are in the R&D stage, with construction targeted for 2020 to 2030. These designs feature advances in sustainability, economics, safety, reliability, and proliferation resistance. Seven of the Generation IV designs are discussed in detail at the WNA site.[218] One intriguing and innovative design by TerraPower would be powered by uranium waste (of which there are 700,000 metric tons in the US alone, waiting to be used or buried). A fleet of these reactors could run for thousands of years on the nuclear waste now available. No mining, enrichment, or reprocessing of fuel would be needed.[219] Funded in part by Bill Gates, through Intellectual Ventures,[220] TerraPower is dedicated to finding zero-emission and sustainable solutions to the climate change crisis. The problem is finding the financial and political backing to get their Generation IV reactor design to become a reality.[221]

Also promising is the Liquid Fluoride Thorium Reactor (LFTR), which may require only 1% of the space of a uranium light-water reactor, use more plentiful thorium fuel at 0.002% of the cost of the uranium fuel, may have a reduced proliferation potential, would be inherently safer, and would produce less than 1% of the waste, with most of the waste having a storage time of 300 years (compared to thousands of years for uranium reactor waste).[222] There are three thorium-related bills pending in Congress right now, including one by Senators Hatch and Reid, and both China and India are considering LFTR designs.[223, 224] However, some of the claims made for the LFTR have been challenged,[225] and huge capital outlays will be required for a completely new reactor design.

Another thorium design being developed by Lightbridge would use a combination of thorium and uranium, but would use essentially the same reactor design as currently in production.[226] The design produces less waste than conventional reactors, and the fuel and its products are unsuitable for weapons development.

Fusion reactors are yet another form of nuclear energy that depend on the energy released by fusing lighter elements into heavier ones — e.g., hydrogen isotopes into helium, as is carried out in the sun's interior. Getting a stable magnetic confinement of the required high-temperature plasma has been difficult to achieve. Efforts at the International Thermonuclear Reactor (ITER) in southern France are leading to testing of the plasma in less than a decade, but with commercial fusion reactors projected to be completed only after 2050.[227] Smaller fusion reactors are being developed as well, which may lead to commercial reactors sooner. One such reactor is the stellerator at the Max Plank Institute for Plasma Physics in Germany, which uses a different magnetic confinement technique than the ITER.[228] Fusion reactors use the isotopes of hydrogen for which there is essentially an unlimited supply in sea water. Furthermore, there are very few radioactive waste products and no danger of a meltdown or theft of bomb-grade materials.

Even with all these promising developments, there are still many challenges that the nuclear industry faces, including safety, proliferation, waste management, and cost.[229]

4. Hydropower

Hydropower makes up about 16% of global electricity (and should comprise about the same share in the year 2035) and 5% to 10% of US electricity, depending on water availability (coming mostly from the western states).[230] In California, hydroelectric power (including small hydro) makes up 15.6% of its total electric power.[231] Hydropower in the US will remain nearly constant over the next 25 years, because most appropriate dam sites have been exploited and any new ones would undoubtedly be opposed by environmental groups. Problems with hydropower include relocation of affected communities, loss of scenic land areas, increase in GHG emissions from decaying biomass in flooded areas, eventual silting up of the lake behind the dam, and lower water volumes expected in many areas where dams could be constructed due to global climate change. So we cannot expect hydropower to pick up much of the renewable power needed in the next century.

5. Biopower

Biomass (materials from recently living plants, as opposed to fossil fuels), especially wood, has been used historically for heating and cooking and still constitutes 10% of the global primary energy consumption. Biopower is biomass-derived electric power, and it constitutes about 1.6% of global electric power production and 1.1% of US production (about one-third of non-hydro renewable electric power).[232] Although biopower is a small part of electricity production today, it is expected to provide 10% of electricity in the US by 2050.[233]

The biomass used in biopower is mostly from wood and wood-derived material, agricultural waste, landfill-produced methane, and municipal solid waste. If the biomass is grown sustainably, it is carbon neutral in the sense that the CO_2 given off during combustion was removed from the atmosphere during its growth phase. It follows, then, that if we bury the CO_2 created from combustion, there would be a net reduction of CO_2 in the atmosphere. Some states and countries have considered this an attractive option in meeting their GHG emission reduction schedules. In addition, biopower is considered base-load power and can be run at full capacity on demand, a valuable complement to intermittent wind or solar power.

Most biopower generation plants are small (less than 50 MW capacity) since they tend to be site specific to avoid the costs of transporting the biomass fuel. For this reason their efficiencies can be only in the low 20% range.[234] When biomass is combined with coal (called *co-firing*) the efficiencies can be as high as those of coal-fired plants, in the 30% to 40% range. Considerably higher efficiencies — over 75% — can be attained in "combined heat and power (CHP)" biopower plants, since the exhaust heat is used for some other electric power cycle or for local industrial purposes or direct space heating.[235] The highest purely electricity-generating biopower plants can achieve efficiencies up to 60%, by creating synthetic gas from the biomass which then fuels a gas turbine (Brayton cycle). The exhaust from the turbine then powers a steam turbine (Rankine cycle). This double-cycle process is called an *integrated gasification combined cycle*. Although still under development, it is the most promising of the biopower plants, especially if carbon capture and sequestration is to be used as well.[236]

An inspiring example that combines all of these ideas (and more) is the Swedish twin cities (Linköping and Norrköping) CTO project for co-generation of electric power, heat, biofuel, and compost.[237] Their project creates a complete ecological cycle, and it even integrates local businesses and jobs.

More recently living plants, complete with their root systems, are being used as extended fuel cells. The electrical power output in test plots has been only about 1.6 watts per square meter, but this is still 7 times more efficient than using the same plants as biomass for biopower generation.[238]

6. Geothermal power

Conventional geothermal power is based on extracting heated water or steam from hydrothermal reservoirs up to 3 km deep. There are two methods for generating electricity. In one known as a *binary cycle plant*, the hot water circulates through a heat exchanger, which in turn heats a fluid that drives a conventional Rankine cycle and generator to produce electricity. In the other type, steam is used to drive a turbine directly, which in turn drives an electrical generator. The binary cycle electricity can be produced for about 5 to 8 cents/kWh, whereas the steam cycle electricity can be produced for only 4 to 6 cents/kWh.[239] In either case, the plants generally run at least 90% of the time, so they make a very reliable base-load renewable power system. As with biopower, this base-load feature is a valuable complement to wind and solar power, which are intermittent. However, the high temperatures and the presence of crystalline rock make drilling into the hydrothermal reservoirs challenging. Much research is dedicated to finding ways to reduce development costs of geothermal energy extraction.

A form of geothermal now under development, known as *enhanced geothermal systems* (EGS), would use the vast hydrothermal potential deeper in the earth's crust, at depths of 10 km. This method is extremely challenging, however, because of the difficulties of drilling so deep and of having to fracture the rock with high-pressure water. But a 2006 study from MIT showed how the necessary technology is available or could be extended from existing capabilities. In their synopsis the authors state: "By evaluating an extensive database of bottomhole temperature and regional geologic data (rock types, stress levels, surface temperatures, etc.), we have estimated the total EGS resource base to be more than 13 million exajoules (EJ) [12.3 million quads]. Using reasonable assumptions regarding how heat would be mined from stimulated EGS reservoirs, we also estimated the extractable portion to exceed 200,000 EJ, or about 2,000 times the annual consumption of primary energy in the United States in 2005. With technology improvements, the economically extractable amount of useful energy could increase by a factor of 10 or more, *thus making EGS sustainable for centuries.*"[240]

In 2010, about 15.7 trillion watt-hours (TWh) of geothermal energy were produced in the US. This is about 0.4% of total US electric energy generation in 2010 and about 10% of non-hydro renewable energy.[241] The US Geological Survey estimates that development of the identified conventional geothermal systems could expand current installed geothermal power by a factor of 3.6, and development of as-yet-undiscovered systems could expand installed power by a factor of 12.[242] With the EIA estimated growth rate for US geothermal power at 4.2% per year,[243] it would be over 60 years before this potential could be fully exploited. If EGS proves feasible, geothermal power could be extracted for centuries.

7. Wind power

The total global wind power capacity at the end of 2009 was about 159 GW (billions of watts), up about 20% from 2008, and it constituted about 2.3% of total global electricity.[244] The US wind capacity increased to 40.2 GW by the end of 2010 (2.9% of total US electricity generation) falling just behind China, which ranked first with 41.2 GW of capacity.[245] The country for which wind generates the highest share of total national electricity is

Denmark, with 19%, a goal the US hopes to reach by 2030.[246] However, a few individual states are close to this; Iowa has reached 16% and South Dakota, 13.6%.[247]

In 2008 the Department of Energy produced a report, in collaboration with US agencies, national research laboratories, and consulting agencies, entitled *20% Wind Energy by 2030: Increasing Wind Energy's Contribution to US Electricity Supply*.[248] The report estimated that 300 GW of total wind-generating capacity would be needed by 2030 in order to produce 20% of the electrical demand at that time. That would mean adding 13 GW of wind capacity every year until 2030. This is not unreasonable, since we added 10 GW in 2009. Their study finds that more than 8000 GW of potential wind-generating capacity can be captured economically in the US. The report also shows where the most appropriate sites are, how the existing power transmission grid would carry the power, and where new power lines would be added.[249]

The investment in reaching the 2030 goal can be supported by measures that favor clean energy and make demands on fossil-fuel-burning energy providers. These include:

- federal and state tax credits,
- Renewable Portfolio Standards (RPS) that require the electric utilities to have a certain percentage of renewable energy sources in the mix of power they deliver, and
- a carbon tax up to $100 per ton of CO_2 produced.

At present, tax incentives have worked well, reducing the cost of wind by 25 %[250] and of solar by up to 50%.[251] RPS have been adopted by 29 US states and the District of Columbia (as of May 2009),[252] and in March of 2011 the California legislature passed a bill mandating 33% renewable electric power by 2020.[253] However, a carbon tax or a cap-and-trade system has not been passed by Congress, and, given the contentious nature of any proposed climate legislation, their enactment seems unlikely for now. Still, the tax incentives, RPS, and ongoing R&D investments and venture capital do seem to be encouraging the development and sale of renewable energy sources. Even now, the price for wind energy in the US is attractive compared to fossil-fuel-generated electricity.[254]

It has been suggested that in order for wind to reach 20% of US supply, wind turbines would have to be sited on off-shore floating platforms. This has the advantage of placing them out of sight from land and also where the winds are more reliable. Such turbines could provide most of the power for the eastern US seaboard and two-thirds of California's needs.[255]

If a substantial part of our energy needs were met by wind, it has been argued that there might be an effect on global climate. Using two general circulation models, Keith et al[256] looked at the effects of running wind turbines in specific arrays at several points worldwide, with an average continuous output of 2000 GW. This is over 40 times the current average global wind power and is about equal to the current total average global electric power output. They found local seasonal temperature fluctuations of about $0.5°C$, but negligible effects on mean global temperature. The wind dissipation outside the array was lower, compensating for the increased dissipation within the array. They suggested that the atmospheric effects of wind power could be mitigated by better turbine design, placing wind towers so the effects cancel, and tailoring the wind farm to the topography so as to minimize added drag.

A major concern with the increasing presence of wind power is the injury to birds. But newer turbine blades are considerably larger and move much slower than the older blades, so they pose little threat to wildlife.

There are three major challenges to reaching the "20% wind energy by 2030" goal:

(a) **The variable nature of wind**
Studies done by the National Renewable Energy Laboratory in 2009 on both the eastern and western US grids determined that there were no fundamental technical barriers to integrating as much as 30% wind-generated electricity into either the Eastern Interconnection or WestConnect

power systems, in spite of the variable nature of wind power. There do need to be frequent scheduling updates and ongoing cooperation between utilities.[257] However, the cost of this kind of management is less than 10% of the value of the wind energy generated.[258] Such cooperation has been ongoing for years in the Danish and northern German electric grids, into which wind energy of up to 30% of total electrical energy demand has been integrated without adding extra reserves.[259] Of particular interest is the balancing between hydro (which can be reduced when wind is strongest) and wind power, since neither produce GHG emissions.

Another method of compensating for intermittent and variable power is to connect uncorrelated wind power sources to the grid (e.g., from different parts of the US[260]), or to connect both wind and solar into the grid. This works for both daily variations (since wind also blows at night) and annual variations (since winds are stronger in winter when sun is at a minimum).[261]

Another use for wind and solar power, independent of their variable natures, is the creation of hydrogen that can be stored for later use in fuel cell vehicles. (Even an airplane has been built that runs on hydrogen.[262])

Other storage ideas for wind energy include pumped hydro, compressed air, flywheels, and various kinds of large battery systems.[263-265] In future decades (when electric cars are more widespread) wind energy can be stored in the batteries of electric and hybrid-electric cars that are connected to the grid, in return for lower electric rates. Alternatively, electric cars can use "smart" chargers that are turned on only when the wind blows.[266]

(b) Expansion of the transmission grid

The Recovery Act of 2009 funded some electric grid extensions (about 6 GW), but up to 270 GW of new extensions are required, if wind is to create 20% of total electric power by 2030.[267] To that end the US government has proposed a modernized grid for the twenty-first century in their *Grid 2030 vision,* which calls for "the construction of a twenty-first-century electric system that connects everyone to abundant, affordable, clean, efficient, and reliable electric power anytime, anywhere."[268] In November of 2010, the Department of Energy announced the funding for five projects aimed at starting this process.[269]

(c) Making wind energy cost competitive

Ten years ago the average wind turbine was out of commission 15% of the time, but now that value is down to 3%.[270] This, along with federal tax incentives, has made a big difference in market acceptance. In addition, the cost of electricity from new wind-power plants is comparable to the cost of electricity from new coal-fired power plants, even without tax incentives (about 10¢ per kWh).[271]

In conclusion, if the pace of new wind-generating capacity continues at about 10% per year, the US will be on track to producing 20% of its electrical generating capacity from wind by 2030.

8. Solar power

Solar energy constitutes only 0.02% of electrical generating capacity in the US (0.06% worldwide),[272] but has grown at 40% to 50% per year for the last 10 years.[273] If the growth of solar can be sustained at 40% per year, it should take about 20 years to reach 20% of generating capacity.

Much of this growth in solar can be met with rooftop solar collectors. For example, California generates 1% of its electrical power from solar (five times the US average), most of it from rooftop collectors. By some estimates, the rooftop potential in California could be 80 times this amount.[274]

But before 20% is reached, the electrical grid will have to adapt to handle the variable nature of both wind and solar (as discussed in the sub-section on wind energy and on the grid sub-section below). Some method of load leveling would be helpful as well. But there is generally a good match between solar output and electrical energy consumption: commercial and industrial use of electricity is highest during the day, when, obviously, solar energy is available.

There are two basic kinds of solar power: (1) solar photovoltaic (PV), which uses semiconductor materials to convert sunlight directly to electricity and constitutes about two-thirds of the installed solar capacity, and (2) concentrated solar power (CSP), which uses the sun's thermal energy to heat a fluid that drives a turbine/generator to create electricity, mostly on a utility-scale basis. The consideration of variability, the grid, and cost mentioned in the wind section above apply to solar energy as well.

(a) Photovoltaic

There are two types of PV cells: the conventional mono/multi-crystalline cells and the newer thin-film deposition cells. The base cost for mono- and multi-crystalline photovoltaic modules ranges between $1.70 and $2.60 per watt (less in large quantities),[275] whereas the thin-film deposition modules are projected to cost below $1 per watt, because far less of the semiconducting material is used in their construction.

Solar PV may enjoy at least three potential advantages over wind energy.

- The cost of some of the thin-film solar collectors may drop well below $1 per watt, a price that wind will find hard to match.
- There are no moving parts, so maintenance costs are minimized.

- Solar PV can be distributed house-to-house, easing the demand for new high-voltage distribution lines.

First Solar, a US-based company, is using a thin-film cadmium-tellurium (Cd-Te) technology in their panels to bring their costs down below $1 per watt.[276] However, extensive use of relatively rare elements like tellurium is not sustainable.[277] As a result, there are efforts to use more common elements like copper, tin, zinc, and sulfur to manufacture solar cells. A project at IBM has resulted in solar cells made from these elements and comparable to the efficiency of commercial cells.[278]

A more radical thin-film technology, which appears to be sustainable, is employed by Nanosolar. According to Nanosolar: "Fundamentally, our cells are a piece of aluminum foil with a micron thick of a copper-based semiconductor on top — all deposited using inexpensive printing processes."[279] They expect their panel price to be as low as $0.60 per watt.

At present, installation costs raise the price per watt for PV dramatically. As noted above, commercial grade conventional crystalline solar cells sell between $1.70 and $2.60 per watt. However, the average *installed* cost in 2009 in the US for a grid-connected PV system before tax incentives was $7.00 per watt (for installations above 1000 kW).[280] Rocky Mountain Institute (RMI) studied installation costs with input from the PV module manufacturers, installers, retailers, and utilities. After considering costs associated with site preparation, structural installation, electrical installation, inverters, and general business processes, they found that it should be possible with efficiencies of scale to get the installation costs down to $0.70 cents per watt.[281] This would bring the total costs for some utility-scale installations down to $1.30 per watt *without subsidy*. At this price the levelized cost (including all maintenance and delivery costs) of solar would be about $0.08 per kWh retail, very competitive with any other source of electrical power.[282] In countries with much lower labor costs than in the US, like China, the cost of PV could be considerably less than any other source of electrical power.

The goal of the renewables industry is to get the price of renewable energy (RE) below that of coal-fired power plants (C), currently the least expensive electrical power. In fact, Larry Page and Sergey Brin of Google have stated that their goal in investing in renewables (such as Nanosolar) is to achieve RE < C without tax incentives; i.e., to get the cost of renewable energy to fall to less than the cost of energy from coal.[283] We may already be approaching the RE < C point, as the analysis above shows. Other analyses predict that, by 2020, solar PV systems will be cost competitive with non-renewable sources of electric power *even without incentives.*[284] Even so, it does seem reasonable to add incentives, such as a carbon tax, to level the playing field, since coal emissions lead to harmful global warming, cause significant health problems, and are responsible for more deaths by far than other forms of electrical energy generation.[285]

For the present, at least, subsidies and tax incentives have helped make solar a larger presence on the grid than it would be otherwise. This is particularly true in China, where the demand for new electricity generation is considerable. Under its "Golden Sun" program, announced in July 2009, the Chinese Ministry of Finance plans to subsidize 50% of the construction costs of grid-connected solar plants;[286] India has similar incentives. The result is that the growth rate of solar generation is projected to be 19% for China and 27% for India.[287] Worldwide, the increase in solar electricity generation was about 50% per year from 2005 to 2008 and about 30% per year thereafter.[288] (However, it is also true that market favoritism can be detrimental to the growth of renewables; in Europe, feed-in tariffs that guarantee above-market rates for solar have led to an overabundance of solar projects, which lowered overall profits for the industry, particularly in Spain.[289)]

Another approach to residential and commercial installations is to lease the solar panels and grid connection device. Currently this is being done in states that offer tax incentives in addition to the 30% federal tax credit.[290] Two of the solar leasing companies operating in California are SolarCity[291] and Sungevity[292]. If the monthly electric bills are high enough (generally over $100/month), the panels can be leased for no money down. The result is that the monthly lease payment plus the average electric bill will cost less than the original average electric bill. As down payments increase, the lease payments decrease. However, leasing costs increase each year, and you don't get to take the federal tax credit, so the total cost with leasing will be more than an outright purchase.[293]

(b) Concentrated solar power (CSP)
About 0.43 GW of CSP was installed by 2008 (less than 5% of PV). CSP is a utility-scale system that uses focused sunlight to heat a fluid which in turn produces steam to drive electricity-generating turbines. Unfortunately, CSP cannot do what PV does, which is operate locally on a small scale where it is needed. Nevertheless, it is receiving both government backing and private investment capital. For example, BrightSource[294] plans to complete their 0.37 GW Ivanpah Solar Energy Generating System in the California Mojave desert by 2013. They have received $1.6 billion in federal loan guarantees and a $168 million investment from Google.[295] The project will be the largest of its kind in the world and will nearly double the US CSP capacity. An advantage of the CSP used at Ivanpah is that the solar energy can be stored in special melted salt pools or in concrete blocks, and released later when needed to generate steam to drive the generators.[296] A further advantage is that the Ivanpah project uses only air to cool the steam in the heat exchangers, as opposed to some

CSP systems that use water — a precious commodity in the desert.[297]

9. Other renewable electrical energy sources

Wave, tidal, and river current power make up very little of the renewable power sources right now. There have been a variety of small demonstration wave and tidal power projects, particularly in northern UK in the Orkney Islands.[298] One intermediate-scale tidal power plant at La Rance, France has been producing an average power of 60 MW since 1966.[299] There have been a number of much larger projects proposed, so wave and tidal power could be a much larger contributor to renewable energy in the future.

10. The smart grid

The fifty-year-old electric power distribution grid in the US is in need of modernization.[300] The "smart grid" is a project conceived on a national scale that would correct most of the current problems. Here are a few of the reasons for pursuing this project:[301]

- To better manage increased electricity demands,
- To avoid brown-outs and black-outs,
- To facilitate the addition of distributed renewable power generation,
- To provide better information for repair crews and minimize down times,
- To delay the need for new power generation,
- To allow automated meter readings,
- To allow individual home power management (to reduce utility bills),

- To provide better management of plug-in hybrid electric vehicles, and
- To lower GHG emissions.

Challenges to achieving this goal include initial cost, lack of standardization, and regulatory barriers. To meet these challenges and advance the smart grid, we must develop national standards, federal funding, investment tax credits, and better consumer awareness.[302] The US government has proposed a modernized grid for the twenty-first century in their *Grid 2030 vision*,[303] and the Department of Energy has announced the funding for five projects aimed at starting this process.[304]

In a commercial venture, Xcel Energy and several vendors have invested $100 million to install two-way "smart meters" in 50,000 homes in Boulder, CO. This technology could reduce peak demand by 27%, thus reducing utility bills and GHG emissions as well. This technology will also make integrating solar and wind energy more practical, and will even allow a plug-in hybrid car battery to be used by the grid for storage in return to lower user rates.[305]

Bibliography

Selected scientific publications

America's Energy Future Panel on Electricity from Renewable Resources. "Electricity from Renewable Resources: Status, Prospects, and Impediments." *National Research Council.* The National Academies Press: 2010. Accessed at http://www.nap.edu/catalog.php?record_id=12619.
(Reviews the status of renewable energies. Also available as a paperback book.)

Hansen, J., et al. "The Case for Young People and Nature: A Path to a Healthy, Natural, Prosperous Future." 2011. Accessed at http://www.columbia.edu/~jeh1/mailings/2011/20110505_CaseForYoungPeople.pdf.
(An up-to-date summary of climate change science, its impacts, and possible corrective scenarios, written by an international team of climate scientists and environmentalists.)

Hansen, J., and M. Sato. "Paleoclimate Implications for Human-Made Climate Change." May 2011. Published on ArXiv.org. Accessed at http://arxiv.org/ftp/arxiv/papers/1105/1105.0968.pdf.
(Based on paleoclimate evidence, this paper argues that global warming must be kept to within 1°C of mid-twentieth-century levels—rather than the 2 to 4°C some have proposed—in order to avoid the worst effects of climate change. There is also an interesting presentation of temperatures during various geological periods.)

Karl, T.R., et al., eds. *Global Climate Change Impacts in the United States (US Global Change Research Program)*. Cambridge University Press, 2009. Accessed at http://www.globalchange.gov/publications/reports/scientific-assessments/us-impacts/full-report .
(A graphic presentation of the likely impacts of climate change expected in the US by the mid and late twenty-first century.)

Lu, X., et al. "Global potential for wind-generated electricity." *Proceedings of the National Academy of Sciences (PNAS)*. 106 no. 27 (Jul 2009): 10933-8, p10937. Accessed at http://www.pnas.org/content/106/27/10933.full.pdf+html.
(A good source for wind power potential worldwide.)

Mann, M.E. "Defining dangerous anthropogenic interference." *PNAS* 106, no. 11 (Mar 2009): 4065-6. Accessed at http://www.pnas.org/content/106/11/4065.full.pdf+html   (18   Jun 2011).

New, M., et al. "Four degrees and beyond: the potential for a global temperature increase of four degrees and its implications." *Phil. Trans. of the Royal Society A,* **369** (Feb 2011): 6-19. Accessed at http://rsta.royalsocietypublishing.org/content/369/1934.toc.
(This article introduces the series [appearing in this issue] on the prospects for a 4°C world and its dangers.)

Pacala, S.W., and R.H. Socolow. "Stabilization Wedges: Solving the Climate Problem for the Next 50 Years with Current Technologies." *Science*, 305 (13 Aug 2004): 968-972. Accessed at   http://www.sciencemag.org/content/305/5686/968   (subscription required).   See also the Carbon Mitigation Initiative at Princeton University accessed at http://cmi.princeton.edu/wedges/pdfs/science_support.pdf.
(This is one of the key papers describing how changes can be made to our energy use that would essentially eliminate carbon emissions.)

Richter, D., and R.A. Houghton. "Gross CO_2 fluxes from land-use change: implications for reducing global emissions and increasing sinks," *Carbon Management* 2, no. 1 (2011). Accessed at http://calhoun.env.duke.edu/files/Richter_DD2011%20Houghton%20gross%20CO2%20sources%20and%20sinks.pdf.
(An excellent resource on the science of land-use-change and its impact on greenhouse gas emissions.)

Smith, J.B., et al. "Assessing dangerous climate change through an update of the Intergovernmental Panel on Climate Change (IPCC): Reasons for concern." *PNAS* 106, no. 11 (17 Mar 2009): 4133-37. Accessed at http://www.pnas.org/content/106/11/4133.full.pdf+html on 18 Jun 2011.

Wiser, R., G. Barbose, and C. Peterman. "Tracking the Sun: The Installed Cost of Photovoltaics in the US from 1998-2007." Lawrence Berkeley National Laboratory, Report No. LBNL-1516E, 2009. Accessed at http://eetd.lbl.gov/ea/ems/reports/lbnl-1516e.pdf.
(A good reference for the economics of solar power.)

Selected governmental, intergovernmental, and agency reports

California Council on Science and Technology (CCST). *California's Energy Future: The View to 2050 – Summary Report,* May 2011. Accessed at http://www.ccst.us/publications/2011/2011energy.pdf.
(From the Introduction: "This report assesses technology requirements for reducing greenhouse gas [GHG] emissions in California to 80% below 1990 levels by 2050 as required by Executive Order S-3-05 2005.")

Committee on Climate Change (CCC). *International Action on Climate Change 2011.* Accessed at http://www.theccc.org.uk/topics/international-action-on-climate-change.
(The CCC was established under the Climate Change Act in 2008 to advise the UK government on issues surrounding climate change.)

Department of Energy (DOE). *20% Wind Energy by 2030: Increasing Wind Energy's Contribution to US Electricity Supply.* July 2008. Accessed at http://www1.eere.energy.gov/windandhydro/pdfs/41869.pdf .

Department of Energy (DOE). *Grid 2030 - A National Vision for Electricity's Second 100 Years.* Accessed at http://www.oe.energy.gov/DocumentsandMedia/Electric_Vision_Document.pdf .

Intergovernmental Panel on Climate Change (IPCC). *Climate Change 2007.* Accessed at http://www.ipcc.ch/publications_and_data/publications_and_data_reports.shtml#1.
(This is the fourth in a series of periodic reports from the international panel of scientists delegated to gather the relevant research results and report on the science, impacts, and mitigation efforts proposed on climate change.)

International Energy Agency (IEA). *Wind Energy Annual Report 2009.* Accessed at http://www.ieawind.org/AnnualReports_PDF/2009/2009AR_92210.pdf.
(The IEA reports on international issues related to energy.)

Center for Climate and Energy Solutions (C2ES). Accessed at http://www.c2es.org (succeeds the Pew Center on Global Climate Change).
(Possibly the best overall assessment of climate change issues and the status of fossil and non-fossil energy use.)

Rocky Mountain Institute. *Energy.* Accessed at http://www.rmi.org/rmi/Energy.

(RMI advocates eliminating fossil fuels as an electrical energy source by 2050, eliminating coal and oil overall by 2050, and reducing natural gas use by one-third.)

US Energy Information Administration (EIA). *Electric Power Monthly*. Accessed at http://www.eia.doe.gov/cneaf/electricity/epm/epm_sum.html.
(Periodic electricity generation data for the US.)

US Energy Information Administration (EIA). *International Energy Outlook 2011*, Sep 2011. Accessed at http://www.eia.gov/forecasts/ieo/.
(The best single source for energy related data by country, region, and worldwide.)

Books

Alley, Richard B. *Earth: the Operator's Manual*. Norton, 2011.
(A good summary of the climate crisis and the technologies that will be competing to avoid the worst impacts of climate change. Richard Alley is one of the world's leading paleoclimatologists and is a member of the IPCC.)

Archer, David, and Stefan Rahmstorf. *The Climate Crisis*. Cambridge UP, 2010.
(A very readable overview of the science and policies of the 2007 IPCC report, by two scientists who contributed to it. They also critique the report, where appropriate, and bring it up to date.)

Brown, Lester. *Plan B 4.0: Mobilizing to Save Civilization*. Norton, 2009.

Brown, Lester. *World on the Edge: How to Prevent Environmental and Economic Collapse*. Norton, 2011.

(Lester Brown has been reviewing the state of the planet for many years and offers some realistic steps we can take to avoid the worst effects of climate change and our over-exploitation of the earth's resources.)

Dauncey, Guy. *The Climate Challenge: 101 Solutions to Global Warming.* New Society Publishers, 2009.
(A good review of climate and energy science and what individuals, local communities, countries, and the global community can do about the climate crisis and needed energy innovations.)

Hansen, James. *Storms of My Grandchildren.* Bloomsbury, 2009.
(A good review of climate science, policy issues, and the struggle to get information out to the public, by one of the world's leading climatologists.)

Letcher, Trevor M., ed. *Climate Change: Observed Impacts on Planet Earth.* Elsevier, 2009.
(A collection of articles by leading scientists on the impacts of climate change.)

MacKay, David J.C. *Sustainable Energy Without the Hot Air.* UIT Cambridge, 2009.
Also available at http://www.inference.phy.cam.ac.uk/sustainable/book/tex/ps/113.252.pdf.
(Physicist MacKay looks at how all of us can reduce our carbon footprint and save money in the process.)

Pittock, A. Barrie. *Climate Change: the Science, Impacts, and Solutions.* CSIRO, 2009.
(Pittock is another contributor to the IPCC reports, who led the Climate Impact Group in the Commonwealth Scientific and Industrial Research Organisation (CSIRO), Australia's national science agency. He discusses the science of climate change from all sides, renewable energy research and technology, and related policies.)

Randers, Jorgen. *2052: A Global Forecast for the Next Forty Years.* Chelsea Green Publishers, 2012.

Intended as a report to the Club of Rome commemorating the 40th anniversary of an earlier book, The Limits to Growth. In this more recent book, Randers updates the projections for growth of population, GDP, energy use, and other parameters to the year 2052, with a focus on sustainability and quality of life.

Smith, Laurence C. *The World in 2050.* Dutton, 2010.

(A good discussion of climate change and its impacts, but Smith also covers the political pressures, geology, ecology, and projected futures of the countries bordering the Arctic as they relate to climate change.)

Selected websites

CleanTech Ostergotland, "Welcome to the Twin Cities of Sweden- Linköping and Norrköping," at http://www.cleante-chostergotland.se/images/CTO_modellen.pdf.

An inspiring example that combines many ideas for sustainable energy and waste management. Their project for co-generation of electric power, heat, biofuel, and compost creates a complete ecological cycle and even integrates local businesses and jobs.

Climate and Energy Report, at http://www.climateandenergyreport.org/

This website has the continually updated pdf online version of this book available free of charge. There is also the opportunity for visitors to the website to leave comments.

Climate Crisis, at http://www.climatecrisis.net/

Features the ideas in *An Inconvenient Truth* (by Al Gore) and suggested actions one can take to reduce one's carbon footprint and alert others to wise energy use.

CO_2Now.org, at http://co₂now.org/.
This organization obtains data from the National Oceanic and Atmospheric Administration (NOAA) and presents it in an easy-to-read fashion.

Earth Policy Institute, at http://www.earth-policy.org.
Covers environmental issues worldwide.

NASA: Global Warming and Climate Change Policy Websites, at http://gcmd.nasa.gov/Resources/pointers/glob warm.html.
Lists some of the websites devoted to climate change and energy technology.

Real Climate, at http://www.realclimate.org/
"A commentary site on climate science by working climate scientists for the interested public and journalists."

Think Progress, at http://thinkprogress.org/tag/global-warming.
Probably the most visited blog on climate change and the political issues pertaining to it - edited by Joe Romm.

350.org, at http://www.350.org/.
This organization is building a global grassroots movement to address the climate crisis. Environmentalist and author Bill McKibben is its principal organizer.

Endnotes

[1] Endnotes referring to websites were downloaded during March and April of 2011, unless otherwise noted.

[2] W.R.L. Anderegg, et al, "Expert Credibility in Climate Change," *Proceedings of the National Academy of Sciences,* June 2010. Accessed at http://www.pnas.org/content/early/2010/06/04/1003187107, on 11 Nov 2011.

[3] Intergovernmental Panel on Climate Change (IPCC), *Climate Change 2007, Working Group I (WG I), Chapter 1: "Historical Overview of Climate Change Science."* Accessed at http://www.ipcc.ch/publications_and_data/ar4/wg1/en/ch1.html .

[4] J. Hansen and M. Sato, "Paleoclimate Implications for Human-Made Climate Change," Jul 2011, published on ArXiv.org. Accessed at http://arxiv.org/abs/1105.0968 on 27 Mar 2012. Also in press in *Climate Change: Inferences from Paleoclimate and Regional Aspects*, Springer.

[5] Ibid., Figures 1 and 2. See also W. Ruddiman, *Earth's Climate*, 2nd ed., (Freeman, 2008), Chapters 9 and 10.

[6] IPCC, *Climate Change 2007, WG I*, Chapter 2: "Changes in Atmospheric Constituents and in Radiative Forcing." Accessed at http://www.ipcc.ch/publications_and_data/ar4/wg1/en/ch2s2-3.html#2-3-1.

[7] From the website CO_2 Now.org at http://co$_2$now.org/.

[8] Hansen and Sato, "Paleoclimate Implications," Figure 2.

[9] This note explains the concept of radiative forcing (RF) in greater detail for the technically inclined. The IPCC defines RF as "a measure of the influence a factor [such as a GHG] has in altering the balance of incoming and outgoing energy in the Earth-atmosphere system and is an index of the importance of the factor as a potential climate change mechanism." (IPCC 2007, Synthesis

Report, Topic 2). That is, certain "factors" (such as GHGs) can serve to increase the greenhouse effect of earth's atmosphere leading to a positive RF. Other factors (such as aerosols) can serve to decrease the greenhouse effect of earth's atmosphere leading to a negative RF. There is also an approximately linear relationship between RF and the corresponding change in global surface temperature in °C: $\Delta T = k \times RF$, where k is approximately 0.8°C per W/m^2. RFs have been calculated for the most important GHGs, including the halocarbons (HC), for the period 1750 to 2005 except for CO_2, methane (CH_4), and nitrous oxide (N_2O), which are updated to 2012 (see Blasing [2012], Endnote 10). These are listed in column 2 in the table below. The RFs for aerosols are negative since these gases reduce the GHG effect (see IPCC 2007, WG1, Chapter 2.9). The lifetimes of the GHG gases and aerosols in the atmosphere are listed in column 3 and their global warming potentials (GWP) over 20-year and 100-year time spans are listed in columns 4 and 5 (see IPCC, *Climate Change 2007, WG I*, Chapter 2.10.2). The GWP approximates the change in RF for a fixed amount of gas relative to the same amount of CO_2 over a specified time span (in this case, 20 and 100 years). So, by definition, CO_2 has a GWP of 1. The GWPs of the other GHGs are all greater than 1.

Gas	RF (W/m²)	Lifetime (years)	GWP 20-yr	GWP 100-yr
CO_2	1.79	100+	1	1
CH_4	0.50	12	72	25
N_2O	0.18	114	289	298
HCs	0.32	10 to 100s	1000s	1000s
Total aerosols	-1.25	hrs to days	—	—

[10] T.J. Blasing, "Recent Greenhouse Gas Concentrations," *Carbon Dioxide Information Analysis Center (CDIAC)*, Feb 2012. Accessed at http://cdiac.ornl.gov/pns/current_ghg.html on 10 Oct 2012.

[11] See table in Endnote 9.

[12] Sometimes emissions are reported in gigatons of CO_2 per year $(GtCO_2/y)$. To convert GtC/y to $GtCO_2/y$, multiply the emissions in Figure 1 by 3.67. This is the ratio of the molecular weight of CO_2 to that of carbon $= 44/12 = 3.67$.

[13] US Energy Information Administration (EIA), "Energy-Related Carbon Dioxide Emissions," *International Energy Outlook 2011.* Accessed at http://www.eia.gov/forecasts/ieo/emissions.cfm on 9 Oct 2012. The cement manufacturing and flaring emissions came from the 2009 CDIAC data set extrapolated to 2010; accessed at http://cdiac.ornl.gov/ftp/ndp030/global.1751_2009.ems on 9 Oct 2012.

[14] Raupach et al., *Proceedings of the National Academy of Sciences (PNAS)*, (12 Jun 2007): 10288, http://www.pnas.org/content/104/24/10288.full.pdf+html .

[15] T. Boden and T.J. Blasing, "Record High 2010 Global Carbon Dioxide Emissions from Fossil-Fuel Combustion and Cement Manufacture" on CDIAC website. Accessed at http://cdiac.ornl.gov/trends/emis/prelim_2009_2010_estimates.html on 27 Mar 2012. See also IEA, "Global Carbon-Dioxide Emissions Increase by 1.0 Gt in 2011 to Record High," 24 May 2012. Accessed at http://www.iea.org/newsroomandevents/news/2012/may/name,27216,en.html on 14 Aug 2012.

[16] D. Richter and R.A. Houghton, "Gross CO_2 Fluxes From Land-Use Change: Implications for Reducing Global Emissions and Increasing Sinks," Carbon Management, 2 (1), 2011, .p2. Accessed at http://calhoun.env.duke.edu/files/Richter_DD2011%20Houghton%20gross%20CO2%20sources%20and%20sinks.pdf on 22 Mar 2012.

[17] Ibid., p3 and Fig.1.

[18] Environmental Protection Agency (EPA), "Global Anthropogenic Non-CO_2 Greenhouse Gas Emissions: 1990-2020," Jun 2006, Table 2-2, accessed at http://www.epa.gov/climatechange/Downloads/EPAactivities/GlobalAnthroEmissionsReport.pdf on 9 Oct 2012. See also IPCC *Climate Change 2007, WG I,* FAQ 7.1.

[19] E.A.G. Schuur and B. Abbott, "High risk of permafrost thaw," *Nature* 480 (1 Dec 2011): 32-33. Accessed at http://www.seas.harvard.edu/climate/eli/Courses/EPS134/Sources/19-Biosphere-feedbacks-amazon-rainforest-and-permafrost/permafrost/Schuur-Abbott-2011_High-risk-of-permafrost-thaw.pdf on 19 Oct 2012. See also Endnote 9.

[20] IPCC, *Climate Change 2007, WG III*, Section 3.3.5.5 "Land Use." Accessed at http://www.ipcc.ch/publications_and_data/ar4/wg3/en/ch3s3-3-5-5.html on 4 Apr 2012.

[21] EPA, "Global Anthropogenic Non-CO_2 Greenhouse Gas Emissions." See also IPCC *Climate Change 2007, WG I*, FAQ 7.1.

[22] EIA, "Energy-Related Carbon Dioxide Emissions," *International Energy Outlook 2011.*

[23] All temperature increases are relative to the global average temperature during mid-twentieth century (or equivalently during the 1951 to 1980 time period) unless stated otherwise. This average temperature is also the maximum average temperature during the Holocene epoch (the last 10,000 years). For more information on temperatures during various geological periods, see J. Hansen and M. Sato, "Paleoclimate Implications for Human-Made Climate Change," May 2011, Figure 1. Accessed at http://arxiv.org/ftp/arxiv/papers/1105/1105.0968.pdf in June 2011.

[24] IPCC, *Climate Change 2007, WG I*, Chapter 8.2: "Advances in Modelling," FAQ 8.1: "How Reliable Are the Models Used to Make Projections of Future Climate Change?" Fig. 1. Accessed at http://www.ipcc.ch/publications_and_data/ar4/wg1/en/faq-8-1.html.

[25] R.A. Muller, *Energy for Future Presidents*, (W.W. Norton, 2012): 41-48. See also IPCC, *Climate Change 2007 WG I*, Chapter 3.2.2: "Temperature in the Instrumental Record for Land and Oceans." Accessed at http://www.ipcc.ch/publications_and_data/ar4/wg1/en/ch3s3-2-2.html.

[26] J.A. Francis and S.J. Vavrus, "Evidence Linking Arctic Amplification to Extreme Weather in Mid-Latitudes," *Geophysical Research Letters* 39, L06801 (2012): 6. Accessed at http://www.agu.org/pubs/crossref/2012/2012GL051000.shtml on 30 Mar 2012. See also J. Gillis, "Weather Runs Hot and Cold, So Scientists Look to the Ice," *New York Times*, 28 Mar 2012.

[27] "Record High Temperatures Far Outpace Record Lows Across US," National Center for Atmospheric Research, Nov 2009. Accessed at https://www2.ucar.edu/atmosnews/news/1036/record-high-temperatures-far-outpace-record-lows-across-us on 30 Mar 2012.

[28] "Interactive Temp Tracker," *Climate Central.* Accessed at www.climatecentral.org/blogs/climate-centrals-record-temperature-tracker/ on 30 Mar 2012. See also K. Trenberth, et al., "Heat Waves and Climate Change," *Climate Communication*, 28 Jun 2012. Accessed at http://climatecommunication.org/wp-content/uploads/2012/06/Heat_Waves_and_Climate_Change.pdf on 10 Jul 2012. See especially Figure 5 of this reference to see how the combined shift and broadening of the probability distribution of temperatures affects the probability of temperature extremes.

[29] IPCC, *Climate Change 2007, WG I*, Chapters 4 and 5. Accessed at http://www.ipcc.ch/publications_and_data/ar4/wg1/en/contents.html.

[30] IPCC, *Climate Change 2007, WG I,* Chapter 6: "Paleoclimate." Accessed at http://www.ipcc.ch/publications_and_data/ar4/wg1/en/ch6.html.

[31] IPCC, *Climate Change 2007, WG I,* Chapter 8: "Climate Models and their Evaluation." Accessed at http://www.ipcc.ch/publications_and_data/ar4/wg1/en/ch8.html (especially FAQ 8.1). For a more thorough discussion of climate models see: J.D. Neelin, *Climate Change and Climate Modeling* (2011).

[32] IPCC, *Climate Change 2007, WGI,* Chapter 8.2, FAQ 8.1. Accessed at http://www.ipcc.ch/publications_and_data/ar4/wg1/en/faq-8-1.html.

[33] T.R. Karl, et al., eds., *Global Climate Change Impacts in the United States (US Global Change Research Program)* (Cambridge University Press, 2009): 22-34. Accessed at http://www.globalchange.gov/publications/reports/scientific-assessments/us-impacts/full-report.

[34] D. Archer, "Methane Hydrate Stability and Anthropogenic Climate Change," *Biogeosciences* 4 (2007):521-4.

[35] E.A.G. Schuur and B. Abbott, "High Risk of Permafrost Thaw," *Nature* 480 (1 Dec 2011): 32-33.

[36] M. Marshall, "Hothouse Earth Is on the Horizon," *New Scientist*, 19 Nov 2011: 10-11.

[37] J.P. Steffensen, et al, "High-Resolution Greenland Ice Core Data Show Abrupt Climate Change Happens in Few Years," *Science* 321, 1 Aug 2008: 680-684. Accessed at http://www.sciencemag.org/content/321/5889/680.full.pdf?sid=d443c36d-5494-4795-ba19-b67e90858328 on 19 Sep 2011. See also Richard B. Alley, *Earth: the Operator's Manual,* (Norton, 2011): 184.

[38] IPCC, *Climate Change 2007,WG II, Climate Change Impacts, Adaptation and Vulnerability.* Accessed at http://www.ipcc.ch/publications_and_data/ar4/wg2/en/contents.html .

[39] IPCC, "Managing the Risks of Extreme Events and Disasters to Advance Climate Change Adaptation," 2012. Accessed at http://s3.documentcloud.org/documents/329332/managing-the-risks-of-extreme-events-and.pdf on 30 Mar 2012.

[40] Center for Climate and Energy Solutions (C2ES) (succeeds Pew Center for Climate Change), *Science and Impacts.* Accessed at http://www.c2es.org on 10 Mar 2012.

[41] T.M. Letcher, ed., *Climate Change: Observed Impacts on Planet Earth,* (Elsevier, 2009).

[42] D. Archer and S. Rahmstorf, *The Climate Crisis: An Introductory Guide to Climate Change*, (Cambridge, 2010).

[43] J.B. Smith, et al, "Assessing Dangerous Climate Change..." *Proceedings of the National Academy of Sciences (PNAS)* 106, no. 11 (17 Mar 2009): 4133-37. Accessed at http://www.pnas.org/content/106/11/4133.full.pdf+html on 18 Jun 2011. See also M.E. Mann, "Defining Dangerous Anthropogenic Interference," *PNAS* 106, no. 11 (17 Mar 2009): 4065-66. Accessed at http://www.pnas.org/content/106/11/4065.full.pdf+html 18 Jun 2011.

[44] Center for Climate and Energy Solutions, *Science and Impacts*. See also endnotes 45 to 73.

[45] P.A. Stott, et al, "Human Contribution to the European Heat Wave of 2003," *Nature* 432 (2004): 610-614.

[46] Karl, *Global Climate Change Impacts in the United States*. See the figure on p 29.

[47] L. Brown, *World on the Edge: How to Prevent Environmental and Economic Collapse*, (Norton, 2011): 3-4.

[48] G. Welton, "The Impact of Russia's 2010 Grain Export Ban," *Oxfam Research Reports*, Jun 2011. Accessed at http://www.oxfam.org/sites/www.oxfam.org/files/rr-impact-russias-grain-export-ban-280611-en.pdf on 4 Apr 2012.

[49] Brown, *World on the Edge*.

[50] S. Rahmstorf, and D. Coumou, "Increase of Extreme Events in a Warming World," *Proceedings of the National Academy of Sciences (PNAS)108*, no. 44 (1 Nov 2011): 17905-09. Accessed at http://www.pnas.org/content/108/44/17905.short on 30 Mar 2012.

[51] G. Luber and M. McGeehin, "Climate Change and Extreme Heat Events," *American Journal of Preventive Medicine 35*, no. 5 (2008): 429–35. Accessed at http://download.journals.elsevierhealth.com/pdfs/journals/0749-3797/PIIS0749379708006867.pdf on 11 Jul 2012.

[52] Richard Leakey, *The Origin of Humankind*, (BasicBooks, 1994): 136.

[53] Hazel Muir, "Thermageddon," *New Scientist*, 23 Oct 2010: 37-9.

[54] M. LePage, "Wild Winters," *New Scientist*, 17 Dec 2011: 42-44. See also Francis and Vavrus, "Evidence Linking Arctic Amplification to Extreme Weather in Mid-Latitudes."

[55] National Weather Service, "May 1 & 2 2010 Epic Flood Event for Western and Middle Tennessee," 18 May 2010. Accessed at http://www.srh.noaa.gov/ohx/?n=may2010epicfloodevent.

[56] Australian Bureau of Meteorology, "Special Climate Statement 24," 7 Jan 2011. Accessed at http://www.bom.gov.au/climate/current/statements/scs24.pdf .

[57] Archer and Rahmstorf, *The Climate Change Crisis*, 136-7.

[58] Karl, *Global Climate Change Impacts in the United States*, 49.

[59] Laurence C. Smith, *The World in 2050*, (Dutton 2010): 101-3.

[60] IPCC, *Climate Change 2007, WG II*, Chapter 13: "Latin America," Box 13.1: "Amazonia: a 'Hotspot' of the Earth System," 604. Accessed at http://www.ipcc.ch/pdf/assessment-report/ar4/wg2/ar4-wg2-chapter13.pdf on 31 Oct 2011.

[61] The National Snow and Ice Data Center, "Arctic Sea Ice Extent Settles At Record Seasonal Minimum," *Arctic Sea Ice News & Analysis*, 19 Sep 2011. Accessed at http://nsidc.org/arcticseaicenews/ on 20 Sep 2012.

[62] M. New, et al., "Four Degrees and Beyond: the Potential for a Global Temperature Increase of Four Degrees and its Implications," *Phil. Trans. of the Royal Society* A, **369 (Feb 2011)**: 6-19. Accessed at http://rsta.royalsocietypublishing.org/content/369/1934.toc .

[63] David Archer, The Long Thaw, (Princeton University Press, 2009): 142.

[64] P. Ward, *The Flooded Earth*, (Basic Books, 2010).

[65] "Ten Years to Save Australia's Great Barrier Reef," *New Scientist*, 9 Apr 2011. Accessed at http://www.newscientist.com/article/mg21028072.500-ten-years-to-save-australias-great-barrier-reef.html .

[66] IPCC, *Climate Change 2007, Synthesis Report*, 54. Accessed at http://www.ipcc.ch/pdf/assessment-report/ar4/syr/ar4_syr.pdf on 11 Sep 2011.

[67] Smith, *The World in 2050*, 5, and also 263 (note 5).

[68] J. Gillis, "With Deaths of Forests, a Loss of Key Climate Protectors," *The New York Times*, 1 Oct 2011. Accessed at http://www.nytimes.com/2011/10/01/science/earth/01forest.html?_r=1&scp=8&sq=rising%20stress%20in%20the%20world%E2%80%99s%20forests&st=cse on 8 Oct 2011.

[69] IPCC, *Climate Change 2007, Synthesis Report*, 48. Accessed at http://www.ipcc.ch/pdf/assessment-report/ar4/syr/ar4_syr.pdf on 11 Sep 2011.

[70] Ibid.

[71] S.L. Pelini, *et al,* "Climate Change and Temporal and Spatial Mismatches in Insect Communities," Chapter 11 in T.M. Letcher, ed., *Climate Change: Observed Impacts on Planet Earth,* (Elsevier, 2009): 227.

[72] Center for Strategic and International Studies, *The Age of Consequences: The Foreign Policy and National Security Implications of Global Climate Change,* 2007. Accessed at http://csis.org/files/media/csis/pubs/071105_ageofconsequences.pdf on 17 Aug 2011.

[73] Center for Naval Analysis, *National Security and the Threat of Climate Change,* 2007. Accessed at http://securityandclimate.cna.org/ on 17 Aug 2011.

[74] The best estimate for *global* temperature increase is based on a climate sensitivity of 3°C for each doubling of CO_2 concentration in the atmosphere. So if the CO_2 concentration doubles from 280 ppm during the Holocene epoch (pre-anthropogenic) to 560 ppm sometime in the future, we would say the best estimate for the temperature rise over that period would be 3°C (relative to the average temperature during the Holocene, which is about equal to the global average temperature in the mid-twentieth century). If the CO_2 concentration doubles again to 1120 ppm, the best estimate for the corresponding temperature increase would be twice that, or 6°C.

Although, to be conservative, we will use the 3°C sensitivity in our calculations, keep in mind that the upper limit of sensitivity that is still considered likely by the IPCC is 4.5°C per doubling (IPCC *Climate Change 2007, WG I,* p 749. Accessed at http://www.ipcc.ch/publications_and_data/ar4/wg1/en/ch10s10-es-1-mean-temperature.html). Furthermore, a more recent analysis finds that a 6°C climate sensitivity (instead of 3°C) is reasonable, if slower surface albedo feedbacks due to vegetation change and melting of the Greenland and Antarctic ice sheets come into play (Previdi et al., "Climate Sensitivity in the Anthropocene," *Earth Syst. Dynam. Discuss.,* 2 (2011): 531–550. Accessed at http://www.earth-syst-dynam-discuss.net/2/531/2011/esdd-2-531-2011.pdf).

[75] IPCC, *Climate Change 2007, WG I,* 749 and 798-9 (Box 10.2).

[76] Ibid.

[77] S.W. Pacala and R.H. Socolow, "Stabilization Wedges: Solving the Climate Problem for the Next 50 Years with Current Technologies," *Science* 305, 13 Aug 2004: 968-972. Accessed at http://www.sciencemag.org/content/305/5686/968. See also supporting on-line material: "The Carbon Mitigation Initiative at Princeton University," Accessed at http://cmi.princeton.edu/wedges/pdfs/science_support.pdf .

[78] Raupach, et al., *Proceedings of the National Academy of Sciences (PNAS)* (12 Jun 2007): 10288. Accessed at http://www.pnas.org/content/104/24/10288.full.pdf+html.

[79] Boden and Blasing, CDIAC website at http://cdiac.ornl.gov/trends/emis/prelim_2009_2010_estimates.html. See also IEA at http://www.iea.org/newsroomandevents/news/2012/may/name,27216,en.html.

[80] CO_2 Now at http://co2now.org/Current-CO2/CO2-Now/global-carbon-emissions.html.

[81] IPCC, *Climate Change 2007, Synthesis Report*, Ch 3.1 "Emissions scenarios," 44.

[82] IPCC, *Climate Change 2007, WG III*, Fig. 3.8, 187. Accessed at http://www.ipcc.ch/pdf/assessment-report/ar4/wg3/ar4-wg3-chapter3.pdf on 20 Mar 2012.

[83] To get the atmospheric CO_2 concentration in the year 2100, we summed the CO_2 emissions from 2010 to 2100, arriving at 2940 GtC (area under the blue curve in Figure 5). This translates into an increase of CO_2 concentration in the atmosphere of about $2940/3.8 = 774$ ppm. The conversion factor 3.8 GtC per ppm is a best estimate, based on 55% of CO_2 emissions going into the atmosphere (the rest into the ocean and land masses), and then each 2.1 GtC of this CO_2 increases the global atmospheric CO_2 concentration by one part per million (ppm). (See IPCC, *Climate Change 2007, WG I*, 824-7. Also Canadell, et al., *PNAS* 104, no. 47 (2007): 18866-70, Fig. 2). Then this CO_2 increase of 774 ppm is added to the CO_2 concentration in 2010 of 390 ppm, giving a total CO_2 concentration in 2100 of 1164 ppm.

To get the temperature rise, we use the best estimate of 3°C per doubling of CO_2 as discussed in the note above. Then the temperature increase, ΔT (relative to mid-twentieth century), can be written as $\Delta T = \log_2([CO_2]/280) \times 3°C = 9.97 \times \log_{10}([CO_2]/280)$ °C, where $[CO_2]$ is the atmospheric concentration of CO_2 in ppm. So in the high-carbon scenario case $\Delta T = 9.97 \times \log_{10}(1164/280)$ or about 6.2°C, with a likely upper estimate of 9.3°C (for the likely upper sensitivity estimate of 4.5°C per doubling).

Note that it may take less than 3.8 GtC of emissions to raise each ppm of atmospheric CO_2 if the oceans slow down their absorption of CO_2 and the earth's forests slow down their absorption because of deforestation, drought, and wildfires. Then our estimates of atmospheric CO_2 and temperature increase at the end of the century would be too low.

[84] J. Hansen and M. Sato, "Paleoclimate Implications for Human-Made Climate Change," published on ArXiv.org ,May 2011, Figure 1. Accessed at http://arxiv.org/ftp/arxiv/papers/1105/1105.0968.pdf. This is a remarkable figure

showing the global temperature history going back 0.5, 5, and 60 million years, in three separate plots.

[85] J. Hansen, et al., "The Case for Young People and Nature: A Path to a Healthy, Natural, Prosperous Future," 2011, p 10. Accessed at http://www.columbia.edu/~jeh1/mailings/2011/20110505_CaseForYoungPeople.pdf. Also IPCC, *Climate Change 2007, WG1,* Chapter 10.7.2, "Climate Change Commitment to Year 3000 and Beyond to Equilibrium." Accessed at http://www.ipcc.ch/publications_and_data/ar4/wg1/en/ch10s10-7-2.html on 25 Oct 2011.

[86] US Energy Information Administration (EIA), *International Energy Outlook 2011,* A2 and A10. Accessed at http://www.eia.gov/forecasts/ieo/more_highlights.cfm#world on 20 Mar 2012.

[87] IPCC, *Climate Change 2007, WGII1,* Chapter 3: 186-187, Fig 3.8.

[88] National Oceanic & Atmospheric Administration (NOAA), "Trends in Atmospheric Carbon Dioxide," Oct 2011. Accessed at http://www.esrl.noaa.gov/gmd/ccgg/trends/ on 25 Nov 2011.

[89] In the moderate-carbon scenario, the emissions increase linearly from 10.56 GtC/y in 2010 to 16.36 GtC/y by 2050, and then increase more gradually to 20 GtC/y by 2100. Similar to the high-carbon scenario, the CO_2 emissions from 2010 to 2100 are summed to get 1450 GtC (area under the red curve in Figure 5). Then the atmospheric concentration of CO_2 in the year 2100 is 390 plus 1450/3.8, or about 770 ppm. The temperature increase by 2100 (relative to mid-twentieth century) is $9.97 \times \log_{10}(770/280) = 4.4°C$, with a likely upper estimate of 6.6°C.

[90] M. New, "Four Degrees and Beyond," op.cit.

[91] Hansen and Sato, "Paleoclimate Implications for Human-Made Climate Change," Figure 6.

[92] Hansen, et al., "The Case for Young People and Nature."

[93] M.Z. Jacobson and M.A. Delucchi, "A Path to Sustainable Energy," *Scientific American,* Nov 2009: 58-65.

[94] Rocky Mountain Institute (RCI), "Energy." Accessed at http://www.rmi.org/rmi/Energy.

[95] H. Lund, *Renewable Energy Systems: The Choice and Modeling of 100% Renewable Solutions,* (Elsevier, 2010).

[96] Pacala and Socolow, "Stabilization Wedges."

[97] Committee on Climate Change, "Building a Low-Carbon Economy — the UK's Contribution to Tackling Climate Change," 1 Dec 2008, especially Chapters 4 through 9. Accessed at http://www.theccc.org.uk/reports/building-a-low-carbon-economy on 19 May 2011.

[98] California Council on Science and Technology (CCST), *California's Energy Future: The View to 2050 Summary Report*, May 2011. Accessed at http://www.ccst.us/publications/2011/2011energy.pdf on 25 May 2011. From the Introduction: "This report assesses technology requirements for reducing greenhouse gas (GHG) emissions in California to 80% below 1990 levels by 2050 as required by Executive Order S-3-05 (2005)."

[99] Hawkins, et al, "What to Do about Coal," *Scientific American*, Sep 2006: 69-75.

[100] California Council on Science and Technology, Rocky Mountain Institute, and Jacobson and Delucchi, op. cit.

[101] IPCC, *Climate Change 2007, Synthesis Report*, Chapter 3.1, p 44.

[102] The total global CO_2 emissions during the twenty-first century for the low-carbon scenario are obtained by calculating the area under the low-carbon curve in Figure 5 between 2000 and 2100. This total is 585 GtC. By comparing this total to 596 GtC, which is the total twenty-first-century emissions for the IPCC SP450 scenario (atmospheric CO_2 stabilizing at 450 ppm), we can estimate that the atmospheric CO_2 concentration for the low-carbon scenario will stabilize at 440 ppm. The reason we did not use the same method as we did for the high- and moderate-carbon scenarios is that, our low-carbon scenario is almost a perfect match for the IPCC SP450 scenario that has been corrected for the slower uptake of CO_2 emissions as the earth experiences global warming. See IPCC, Climate Change 2007, WG I, Chapter 10.4.1, accessed at http://www.ipcc.ch/publications_and_data/ar4/wg1/en/ch10s10-4.html on 2 Aug 2011.

The IPCC SP450 temperature rise is 2.1°C by the year 2100 relative to temperatures in the first half of the twentieth century. This is equivalent to a 1.95°C rise relative to mid-twentieth century temperatures, which is the baseline for figuring temperature increases in this report.

[103] To lower the atmospheric CO_2 concentration to 350 ppm would require replacing even natural-gas-fired power plants with renewable or nuclear power and getting part of our transportation energy from hydrogen, which could be made at wind or solar power plants. This could lower the atmospheric CO_2 concentration to the range of 350 ppm to 370 ppm by the end of the century.

[104] G. Stoyke, *The Carbon Buster's Home Energy Handbook*, (New Society Publ., 2007).

G. Dauncey, *The Climate Challenge: 101 Solutions to Global Warming,* (New Society Publ., 2009).

David J.C. MacKay, *Sustainable Energy Without the Hot Air,* (UIT Cambridge, 2009). Also available at http://www.inference.phy.cam.ac.uk/sustainable/book/tex/ps/113.252.pdf.

Energy Star, US Environmental Protection Agency, http://www.energystar.gov.

"Energy Saving Tips for Your Home or Office," *Get Energy Smart.* Accessed at http://www.ccicenter.org/GetEnergySmarter/EnergySavingTips.aspx on 31 Mar 2012.

PG&E, "Energy Saving Tips." Accessed at http://www.pge.com/myhome/saveenergymoney/savingstips on 31 Mar 2012.

[105] CCST, *California's Energy Future,* op. cit.

[106] As an example, a typical small-room natural-gas heater generates 16,000 BTU/hr and requires an input of 24,000 BTU/hr (since a third of the energy is lost in venting), costing about 24 cents/hr. A Fujitsu 12RLS heat pump (http://www.fujitsugeneral.com/PDF_06/halcyon06_brochure.pdf p12) requires 1.2 kWh/hr (4094 BTU equivalent) to heat at the same 16,000 BTU/hr rate, at a cost of about 12 cents/hr—half as much as the gas heater. The same heater generates about 0.1345 lb of CO_2/1000 BTU (http://energyexperts.org/EnergySolutionsDatabase/ResourceDetail.aspx?id=4947), or 3.23 lb of CO_2/hr. For the average US electric power plant, the CO_2 emissions are 1,34 lb of CO_2 per kWh (ftp://ftp.eia.doe.gov/pub/oiaf/1605/cdrom/pdf/e-supdoc.pdf Table 1), or 1.61 lb of CO_2/hr for the heat pump, or about half the CO_2 emissions of the gas heater. For the US Pacific states, the heat pump would be responsible for only one-sixth the CO_2 emissions of the gas heater.

[107] Research and Innovative Technology Administration (RITA), Tables 4-6, 4-11, 4-12, and 4-23: *Average Fuel Efficiency of US Passenger Cars and Light Trucks.* Accessed at http://www.bts.gov/publications/national_transportation_statistics/ on 1 Apr 2012.

[108] This assumes the car is used only in electric mode for a hybrid, all driving is done within range of a charging station, and the cost of electricity is 10 cents/kWh. (This cost can be significantly less with EV charging rates.)

Costs are calculated for the Nissan Leaf, but would be similar for the Coda. The full charge on a Leaf is 24 kWh (kilowatt-hours) for a range of 100 miles, so, for the annual distance of 11,450 miles, the energy used is 2750 kWh and the cost is $275.

[109] The most expensive component for maintenance would be the battery, but it is guaranteed for 100,000 miles or 8 years. According to Nissan it should still have 70 to 80% of its life still left after that, so it would make sense to replace individual modules in the battery at $600 apiece, if needed.

Energy Efficiency and Renewable Energy, "How Do Gasoline and Electric Vehicles Compare?" Department of Energy. Accessed at http://www1.eere.energy.gov/vehiclesandfuels/avta/light_duty/fsev/fsev_gas_elec1.html on 6 Mar 2012. Also, "10 Things You (Probably) Didn't Know About the Nissan Leaf," *Popular Mechanics*. Accessed at http://www.popularmechanics.com/cars/reviews/hybrid-electric/10-things-you-didnt-know-about-the-nissan-leaf on 11 Mar 2012.

[110] A gallon of gasoline generates 19.66 lb of CO_2, so a conventional car getting 22.6 mpg generates 0.87 lb of CO_2 per mile. The Nissan Leaf uses 0.24 kWh per mile, but with 20% transmission and charging losses included, this is raised to 0.30 kWh per mile. So for the average US electric power plant (1.34 lb of CO_2 per kWh), the Leaf generates 0.40 lb of CO_2 per mile, only 46% of the CO_2 of the conventional car. For US Pacific states (0.45 lb of CO_2 per kWh), the Leaf generates only 16% of the CO_2 of the conventional car. See Clean Energy at http://www.epa.gov/cleanenergy/energy-resources/refs.html, and the EIA, "Updated State-level Greenhouse Gas Emission Coefficients for Electricity Generation," at ftp://ftp.eia.doe.gov/pub/oiaf/1605/cdrom/pdf/e-supdoc.pdf.

[111] "Why Yes, a Fully Charged Nissan Leaf Can Go 87 Miles with Range to Spare," *Motor Trend*, 25 Apr 2011. Accessed at http://blogs.motortrend.com/fully-charged-nissan-leaf-87-miles-range-spare-14485.html on 8 Jun 2011.

Also, D.R. Baker, "Coda Electric Sedan Rolls off Benicia Assembly Line," *San Francisco Chronicle*, 13 Mar 2012. Accessed at http://www.sfgate.com/cgi-bin/article.cgi?f=/c/a/2012/03/13/BUP01NJNS4.DTL on 13 Mar 2012.

[112] M. Fischetti, "Can US Cars Meet the New 54 mpg CAFE Standards? Yes They Can," *Scientific American*, 16 Nov 2011. Accessed at http://blogs.scientificamerican.com/observations/2011/11/16/can-cars-meet-the-new-54-mpg-cafe-standards-yes-they-can/ on 1 Apr 2012.

[113] Research and Innovative Technology Administration (RITA), Tables 4-6, 4-11, 4-12, and 4-23: *Average Fuel Efficiency of US Passenger Cars and Light Trucks*. Accessed at http://www.bts.gov/publications/national_transportation_statistics/ on 1 Apr 2012. Using these tables, one finds that, by switching cars and light trucks from their current average of 20.7 mpg to 50 mpg by 2050 (consistent with new federal standards for 2025), we would save over 9 quads of energy (one quad = one quadrillion BTU = 1000 trillion BTU.) Making half of those vehicles electric cars or plug-in hybrids saves another 3 quads (after

electric utility consumption is figured in). Fuel savings on half of the heavy trucks and buses saves another 3 quads, and one-third savings on the rest of the transportation sector saves another 1.5 quads. The savings total 16.5 quads of energy from oil, which is about two-thirds of the total US transportation sector energy use (25 quads, almost entirely due to petroleum products).

This analysis considered only the cut in oil consumption in the transportation sector (a two-thirds cut). If the emissions from electricity generation needed to run the electric vehicles are included, we find conservatively that there is still a net 50% reduction in CO_2 emissions in the transportation sector. If the US can make this transformation by 2050 (admittedly a tough job), then Europe, China, and India (half of the rest of the world's population) can make the transformation, since they are already more invested in small energy-efficient vehicles.

[114] Center for Climate and Energy Solutions (C2ES), *Climate Techbook/ Transportation/ Biofuels Overview.* Accessed at http://www.c2es.org/ technology/overview/biofuels on 13 Mar 2012.

[115] Center for Climate and Energy Solutions, *Climate Techbook/Transportation/ Cellulosic Ethanol,* op. cit.

[116] "Building a Brighter Future: California's Progress Toward a Million Solar Roofs," Environment California. Accessed at http://www. environmentcalifornia.org/reports/cae/building-brighter-future-california%E2%80%99s-progress-toward-million-solar-roofs on 11 Mar 2012.

[117] G. Barbose, et al., *Tracking the Sun III: The Installed Cost of Photovoltaics in the US from 1998-2009*, Dec 2010. Accessed at http://eetd.lbl.gov/ea/ems/ reports/lbnl-4121e.pdf on 8 Apr 2011.

[118] Rocky Mountain Institute, "Achieving Low-Cost Solar PV: Industry Workshop Recommendations for Near-Term Balance of System Cost Reductions." Accessed at http://www.rmi.org/Content/Files/BOSReport. pdf on 7 Jun 2011.

[119] Sun Electronics at http://www.sunelec.com/index.php?main_page=san_ francisco .

[120] P.R. Epstein, "Full Cost Accounting for the Life Cycle of Coal," *Annals of the New York Academy of Sciences* 1219, Feb 2011: 73. Accessed at http://onlinelibrary.wiley.com/doi/10.1111/j.1749-6632.2010.05890.x/full on 19 Jul 2011.

[121] L. Brown, *Plan B 4.0: Mobilizing to Save Civilization*, Norton (2009).

[122] J. Hansen, *Storms of My Grandchildren*, Bloomsbury (2009): 209.

[123] Sandler and DeShazo, "Carbon Costs," UCLA, 2008. Accessed at http://lewis.spa.ucla.edu/publications/reports/Sandler_Deshazo_Climate.pdf on 18 May 2011.

[124] Hansen, *Storms of My Grandchildren.*

[125] S. Stoft, *Carbonomics*, (Diamond, 2008): 108.

[126] "An Expensive Gamble," *The Economist*, 16 Jul 2011: 41.

[127] A. Rourke, "Australian MPs Pass Carbon Tax," *The Guardian*, 12 Oct 2011, accessed at http://www.guardian.co.uk/environment/2011/oct/12/australian-carbon-tax-passed on 4 Oct 2012. "Australian Senate Passes Carbon Tax," *The Guardian*, 7 Nov 2011, accessed at http://www.guardian.co.uk/world/2011/nov/08/australia-senate-passes-carbon-tax on 4 Oct 2012. P. Martin, "Figures Show Inflation Effect of Carbon Tax Mostly Hot Air," *The Sydney Morning Herald*, 7 Aug 2012, accessed at http://www.smh.com.au/opinion/political-news/figures-show-inflation-effect-of-carbon-tax-mostly-hot-air-20120806-23qd8.html on 4 Oct 2012..

[128] S. Haszeldine and V. Scott, "Carbon Capture and Storage," *New Scientist*, 12 Apr 2011. See also E. Mathez, *Climate Change*, Columbia (2010): 193-197.

[129] P. Byck, "Green-Spangled Banner," *New Scientist*, 28 May 2011: 26-7.

[130] F. Barringer, "California Adopts Limits on Greenhouse Gases," *The New York Times*, 20 Oct 2011: A25. Accessed at http://www.nytimes.com/2011/10/21/business/energy-environment/california-adopts-cap-and-trade-system-to-limit-emissions.html?_r=1&scp=1&sq=california%20adopts%20limits&st=cse on 22 Oct 2011.

[131] L. Brown, *US Moving Toward Ban on New Coal-Fired Power Plants*, Earth Policy Institute, 14 Feb 2008. Accessed at http://www.earth-policy.org/index.php?/plan_b_updates/2008/update70 .

[132] Ibid.

[133] R. Smith, "The Coal Age Nears its End," *The Wall Street Journal*, 23 Dec 2011: B1.

[134] Center for Climate and Energy Solutions (C2ES), *Climate Techbook/Electricity/Solar Power*. Accessed at http://www.c2es.org/technology/factsheet/solar on 13 Mar 2012.

[135] D.R. Baker, "Renewable-Energy Bill Passed by Legislature," *San Francisco Chronicle*, 30 Mar 2011. Accessed at http://www.sfgate.com/cgi-bin/article.cgi?f=/c/a/2011/03/30/MN4L1IM38G.DTL .

[136] Energy Information Administration (EIA), "US Energy-Related CO_2 Emissions in Early 2012 Lowest Since 1992," *Today in Energy*, 1 Aug 2012. Accessed at http://www.eia.gov/todayinenergy/detail.cfm?id=7350 on 19 Aug 2012.

[137] R. Nuwer, "A 20-Year Low in US Carbon Emissions," *New York Times*, 17 Aug 2012. Accessed at http://green.blogs.nytimes.com/2012/08/17/a-20-year-low-in-u-s-carbon-emissions/ on 19 Aug 2012.

[138] Center for Climate and Energy Solutions, *Climate Techbook/Electricity/Biopower.*

[139] "Substantial Power Generation from Domestic Geothermal Resources," *USGS Newsroom*, US Geological Survey, 29 Sep 2008. Accessed at http://www.usgs.gov/newsroom/article.asp?ID=2027 on 8 Apr 2012.

[140] Jefferson Tester, et. al. "The Future of Geothermal Energy: Impact of Enhanced Geothermal Systems (EGS) on the United States in the 21st Century." Massachusetts Institute of Technology, 2006. Accessed at http://www1.eere.energy.gov/geothermal/pdfs/future_geo_energy.pdf on 3 Apr 2012.

[141] Richter and Houghton, "Gross CO_2 Fluxes."

[142] Ibid., 3 and Fig.1.

[143] IPCC, *Climate Change 2007, WG III*, Chapter 9.6.1: "Policies aimed at reducing deforestation." Accessed at http://www.ipcc.ch/publications_and_data/ar4/wg3/en/ch9s9-6-1.html .

[144] Richter and Houghton, "Gross CO_2 Fluxes."

[145] J.R. Teasdale, "Strategies for Soil Conservation in No-Tillage and Organic Farming Systems," *Journal of Soil and Water Conservation* 62, no. 6 (2007). Accessed at http://afrsweb.usda.gov/SP2UserFiles/Place/12650400/JSWC200762-6-144-147.pdf on 13 Mar 2012.

[146] Brown, *World on the Edge.*

[147] Teasdale, "Strategies for Soil Conservation."

[148] F. Pearce, "Methane Cuts Could Delay Climate Change by 15 years," *New Scientist*, 31 Mar 2012: 10-11. Accessed at http://www.newscientist.com/article/mg21328583.800-methane-cuts-could-delay-climate-change-by-15-years.html on 15 Apr 2012.

[149] California Department of Resources Recycling and Recovery, "Landfill Methane Capture Strategy." Accessed at http://www.calrecycle.ca.gov/climate/Landfills/default.htm on 13 Mar 2012.

[150] IPCC, *Climate Change 2007, WG III*, Section 3.3.5.5: "*Land Use.*" Accessed at http://www.ipcc.ch/publications_and_data/ar4/wg3/en/ch3s3-3-5-5.html on 4 Apr 2012. Also, Agricultural Research Service, USDA, "Improved Measurement, Monitoring, and Mitigation of Nitrous Oxide Emissions and Related N Losses from Intensively Fertilized Agro-Ecosystems." Accessed at http://www.ars.usda.gov/research/projects/projects.htm?accn_no=415081 on 4 Apr 2012.

[151] IPCC, *Climate Change 2007, WG III*, 55-7. Accessed at http://www.ipcc.ch/pdf/assessment-report/ar4/wg3/ar4-wg3-ts.pdf on 4 Apr 2012.

[152] United Nations Department of Economic and Social Affairs, "Renewable Energy." Accessed at http://www.un.org/en/development/desa/climate-change/renewable-energy.shtml on 12 Mar 2012.

[153] P. Runci, "Renewable Energy Policy in Germany." Joint Global Change Research Institute. Accessed at http://www.globalchange.umd.edu/energytrends/germany/ on 12 Mar 2012.

[154] National Renewable Energy Laboratory (NREL) website, accessed at http://www.nrel.gov/ on 12 Mar 2012 and The Office of Energy Efficiency and Renewable Energy (EERE) website, accessed at http://www.eere.energy.gov/.

[155] US Department of State, *U.S. Climate Action Report 2010*, Chapter 7, Global Publishing Services (Jun 2010), accessed at http://www.state.gov/documents/organization/140005.pdf on 3 Oct 2012.

UN Environment Program, "Financing Renewable Energy in Developing Countries," UNEP Finance Initiative (Feb 2012), accessed at http://www.unepfi.org/fileadmin/documents/Financing_Renewable_Energy_in_subSaharan_Africa.pdf on 3 Oct 2012.

Organization for Economic Cooperation and Development (OECD), "Financing Climate Change Action and Boosting Technology Change," (2012), accessed at http://www.oecd.org/environment/climatechange/46534686.pdf on 3 Oct 2012.

[156] G. van de Kerk and A.R. Manuel, "Sustainable Society Index," *The Encyclopedia of the Earth*. Accessed at http://www.eoearth.org/article/Sustainable_Society_Index on 17 Mar 2012.

[157] Brown, *World on the Edge.*

[158] World Footprint, *Global Footprint Network.* Accessed at http://www.footprintnetwork.org/en/index.php/GFN/page/world_footprint/ on 17 Mar 2012.

[159] "China GDP Annual Growth Rate," *TradingEconomics.* Accessed at http://www.tradingeconomics.com/china/gdp-growth-annual on 29 Mar 2012.

[160] US Energy Information Administration (EIA), *International Energy Outlook 2011*, A7. Accessed at http://www.eia.gov/forecasts/ieo/more_highlights.cfm#world on 20 Mar 2012.

[161] R. Heinberg, *The End of Growth*, (New Society Publishers, 2011): 192.

[162] J. Diamond, *Collapse: How Societies Choose to Fail or Succeed*, (Penguin Books, 2005).

[163] D. Meadows, et al, *Limits to Growth*, (Chelsea Green Publishing, 2004).

[164] Brown, *Plan B 4.0* and *World on the Edge.*

[165] C. Martenson, *The Crash Course: The Unsustainable Future Of Our Economy, Energy, And Environment*, (Wiley, 2011).

[166] J. Sachs, *The Price of Civilization*, (Random House, 2011).

[167] "The 9 billion-people question—A special report on feeding the world," *The Economist*, 26 Feb 2011.

[168] "7 billion and counting," *New Scientist*, 26 Sep 2009: 35-44.

[169] See also Section III.B.2 and Appendix B (Section B.1) of this work.

[170] See notes in Section III.A.3 and 4 of this work.

[171] CCST, *California's Energy Future.*

[172] There are no operating CO_2 emissions, but there are emissions associated with the construction and decommissioning of the nuclear plant and the mining and transport of uranium. See K. Kleiner, "Nuclear Energy: Assessing the Emissions," *Nature Reports Climate Change*, 24 Sep 2008. Accessed at www.nature.com/climate/2008/0810/full/climate.2008.99.html on 18 May 2012.

[173] There is a small amount of emissions from the manufacture of renewable systems. For more details on renewable electricity, see Section III.B.2 and Appendix B (Sections B4 through B10) of this work.

[174] Hendriks, et al, "Emission Reduction of Greenhouse Gases from the Cement Industry," *International Conference on Greenhouse Gas Technologies*, 8 Apr 2012.

Accessed at http://www.wbcsd.org/web/projects/cement/tf1/prghgt42.pdf on 8 Apr 2012.

[175] G. Dauncey, *The Climate Challenge: 101 Solutions to Global Warming*, New Society, (2009): 38-39. See also "Carbon Dioxide Emissions," US Environmental Protection Agency accessed at http://www.epa.gov/climatechange/ghgemissions/gases/co$_2$.html on 15 Dec 2012.

[176] Research and Innovative Technology Administration (RITA), Table 4-23: *Average Fuel Efficiency of US Passenger Cars and Light Trucks*. Accessed at http://www.bts.gov/publications/national_transportation_statistics/html/table_04_23.html .

[177] RITA, Table 4-9: *Motor Vehicle Fuel Consumption and Travel*. Accessed at http://www.bts.gov/publications/national_transportation_statistics/html/table_04_09.html . Also EIA, *International Energy Outlook 2011*, A2.

[178] Energy Efficiency and Renewable Energy, "How Do Gasoline and Electric Vehicles Compare?" Department of Energy. Accessed at http://www1.eere.energy.gov/vehiclesandfuels/avta/light_duty/fsev/fsev_gas_elec1.html on 6 Mar 2012. Also, "10 Things You (Probably) Didn't Know About the Nissan Leaf," *Popular Mechanics*. Accessed at http://www.popularmechanics.com/cars/reviews/hybrid-electric/10-things-you-didnt-know-about-the-nissan-leaf on 11 Mar 2012.

[179] "No Rare Earth Metals in the Model S," at http://www.teslamotors.com/forum/forums/no-rare-earth-metals-model-s /

[180] E. Grabianowski, "How the Tesla Roadster Works," at http://auto.howstuffworks.com/tesla-roadster.htm/printable .

[181] "A New Process Will Make Solid-State Rechargeable Batteries That Should Greatly Outperform Existing Ones" *The Economist*, 27 Jan 2011. Accessed at http://www.economist.com/node/18007516 .

[182] B. Liggett, "New Lithium-Air Batteries Could Go 500 Miles on a Single Charge," *Inhabitat*, 11 Apr 2011. Accessed at http://inhabitat.com/new-lithium-air-batteries-could-go-500-miles-on-a-single-charge/ .

[183] M. LaMonica, "Ultracapacitors Look to Fit into Energy Storage," *Green Tech*, 29 Sep 2009. Accessed at http://news.cnet.com/8301-11128_3-10363496-54.html .

[184] Center for Climate and Energy Solutions, *Climate Techbook/Transportation/Hydrogen fuel cell vehicles*. Accessed at http://www.c2es.org/climate-techbook on 10 Mar 2012.

[185] Center for Climate and Energy Solutions, *Climate Techbook/Transportation/ Biofuels Overview.*

[186] David Bransby, "Switch Grass: Alternative Energy Source?" Appearance on National Public Radio, 1 Feb 2006. Accessed at http://www.npr.org/ templates/story/story.php?storyId=5183608 .

[187] M. Wang, et al., "Life-cycle Energy and Greenhouse Gas Emission Impacts of Different Corn Ethanol Plant Types," *Environmental Research Letters* 2, 2007: 1-13.

[188] Center for Climate and Energy Solutions, *Climate Techbook/Transportation/ Cellulosic Ethanol.*

[189] Center for Climate and Energy Solutions, *Climate Techbook/Transportation/ Ethanol.*

[190] International Energy Agency (IEA), "Biofuels Can Provide up to 27% of World Transportation Fuel by 2050," 20 Apr 2011. Accessed at http://www.iea. org/press/pressdetail.asp?PRESS_REL_ID=411 on 17 Jun 2011.

[191] Federal Ministry for the Environment, *Development of renewable energy sources in Germany (2010):* 3. Accessed at http://www.erneuerbare-energien.de/files/ english/pdf/application/pdf/ee_in_deutschland_graf_tab_en.pdf on 20 Oct 2011.

[192] California Energy Commission, *Total Electricity System Power* (2010). Accessed at http://energyalmanac.ca.gov/electricity/total_system_power.html on 28 Sep 2012.

[193] Energy Information Administration (EIA), "Levelized Cost of New Generation Resources in the Annual Energy Outlook 2012," 25 Jun 2012. Accessed at http://www.eia.gov/forecasts/aeo/electricity_generation.cfm on 4 Oct 2012.

[194] EIA, *International Energy Statistics, Electricity.* Accessed at http://www.eia. gov/cfapps/ipdbproject/IEDIndex3.cfm# on 3 Aug 2011. For US electricity generation in 2011, see EIA, "What is US Electricity Generation by Energy Source?" *Frequently Asked Questions.* Accessed at http://www.eia.gov/tools/ faqs/faq.cfm?id=427&t=3 on 17 Aug 2012.

[195] Center for Climate and Energy Solutions, *Techbook/Electricity/Solar Power.*

[196] For details see Sections B.7 and B.8 on wind and solar in this appendix.

[197] "Award-winning Solar Powered Desalination Unit aims to solve Central Australian water problems," University of Wollongong, 4 Nov 2005. Accessed at http://media.uow.edu.au/news/2005/1104c/index.html on 28 Oct 2011.

P. Patel, "Solar-Powered Desalination," *Technology Review* (MIT), 8 Apr 2010. Accessed at http://www.technologyreview.com/energy/25010/ on 28 Oct 2011.

[198] M. Wines, "China Takes a Loss to Get Ahead in the Business of Fresh Water," *The New York Times*, 26 Oct 2011.

[199] EIA, *International Energy Statistics, Electricity*. Accessed at http://www.eia.gov/cfapps/ipdbproject/IEDIndex3.cfm# on 3 Aug 2011.

[200] EIA, "US Energy-Related CO_2 Emissions in Early 2012 Lowest Since 1992," *Today in Energy*, 1 Aug 2012. Accessed at http://www.eia.gov/todayinenergy/detail.cfm?id=7350 on 19 Aug 2012. See also EIA, *International Energy Outlook 2011*, A10.

[201] S. Ansolabehere, et al., "The Future of Coal," MIT, 2007. Accessed at http://web.mit.edu/coal/ . Also in Stoft, "Carbonomics," 108.

[202] W. Broecker and R. Kunzig, *Fixing Climate*, (Hill & Wang, 2008): 198-225.

[203] J. Fallows, "Dirty Coal, Clean Future," *Atlantic Monthly*, Dec 2010. Accessed at http://www.theatlantic.com/magazine/archive/2010/12/dirty-coal-clean-future/8307/ on 20 May 2011.

[204] "Carbon Capture and Storage: A Shiny New Pipe Dream," *The Economist*, 12 May 2012. Accessed at http://www.economist.com/node/21554501 on 19 May 2012.

[205] Haszeldine and Scott, "Carbon Capture and Storage."

[206] IPCC, *Climate Change 2007, WG III*, Chapter 4.3.6: "*Carbon Dioxide Capture and Storage (CCS).*" Accessed at http://www.ipcc.ch/publications_and_data/ar4/wg3/en/ch4s4-3-6.html .

[207] A.B. Pittock, *Climate Change: the Science, Impacts, and Solutions*, CSIRO, 2009: 190-194.

[208] R.B. Alley, *Earth: the Operator's Manual*, (Norton & Co, 2011): 285-290.

[209] EIA, *Annual Energy Outlook 2011 Early Release Overview*, Figure 12. Accessed at http://www.eia.gov/forecasts/aeo/pdf/0383er%282011%29.pdf .

[210] EIA, *International Energy Outlook 2011*, A1 and A2. Accessed at http://www.eia.gov/forecasts/ieo/more_highlights.cfm#world on 23 Mar 2012.

[211] EIA, *Fuel Emission Coefficients*, 2011. Accessed at http://www.eia.doe.gov/oiaf/1605/coefficients.html .

[212] "German Energy: Nuclear? Nein, danke," *The Economist*, 4 Jun 2011.

[213] "Nuclear Power: When the Steam Clears," *The Economist*, 26 Mar 2011. Accessed at http://www.economist.com/node/18441163 on 17 Oct 2011. A less optimistic view was presented a year later: "The Dream That Failed," *The Economist*, 10 Mar 2012.

[214] K. Kleiner, "Nuclear Energy: Assessing the Emissions," *Nature Reports Climate Change*, 24 Sep 2008. Accessed at www.nature.com/climate/2008/0810/full/climate.2008.99.html on 18 May 2012

[215] P. McKenna, "Coal is Far Deadlier Than Nuclear Power," *New Scientist*, 26 Mar 2011.

[216] P.R. Epstein, "Full Cost Accounting for the Life Cycle of Coal," *Annals of the New York Academy of Sciences*, 1219 (Feb 2011): 73. Accessed at http://onlinelibrary.wiley.com/doi/10.1111/j.1749-6632.2010.05890.x/full on 19 Jul 2011.

[217] World Nuclear Association, *Advanced Nuclear Power Reactors*. Accessed at http://www.world-nuclear.org/info/inf08.html .

[218] World Nuclear Association, *Generation IV Nuclear Reactors*. Accessed at http://www.world-nuclear.org/info/inf77.html.

[219] TerraPower, *Traveling Wave Reactor*. Accessed at http://www.terrapower.com/Home.aspx .

[220] Intellectual Ventures. Accessed at http://www.intellectualventures.com .

[221] R.A. Guth, "*A Window into the Nuclear Future*," *The Wall Street Journal*. Accessed at http://online.wsj.com/article/SB10001424052748704409004576146061231899264.html .

[222] R. Hargraves at https://docs.google.com/viewer?url=http%3A%2F%2Fhome.comcast.net%2F~robert.hargraves%2Fpublic_html%2FAimHigh.pdf

[223] J. Quinn, "Safe Nuclear Does Exist, and China Is Leading the Way with Thorium," *The Telegraph*, 11 Apr 2011. Accessed at http://www.telegraph.co.uk/finance/comment/ambroseevans_pritchard/8393984/Safe-nuclear-does-exist-and-China-is-leading-the-way-with-thorium.html in Apr 2011.

[224] R. Martin, "Uranium Is So Last Century—Enter Thorium, the New Green Nuke," *Wired Magazine*, Jan 2010. Accessed at http://www.wired.com/magazine/2009/12/ff_new_nukes/all/1 in Mar 2011.

[225] D.A. Ryan, "The Molten Salt Reactor Concept," Part 8 of "*A Critical Analysis of Current and Proposed Future Nuclear Reactors Designs*," on *daryanenergyblog*. Accessed at http://daryanenergyblog.wordpress.com/ca/ on 8 Oct 2011.

See also A. Makhijani and M. Boyd, "Thorium Fuel: No Panacea for Nuclear Power," Institute for Energy and Environmental Research and Physicians for Social Responsibility. Accessed at http://www.ieer.org/fctsheet/thorium2009factsheet.pdf on 8 Oct 2011

[226] Makhijani and Boyd, "Thorium Fuel."

[227] Facts and Figures at ITER, at http://www.iter.org/factsfigures . Accessed on 21 Sep 2011.

[228] "Fusion power," *The Economist*, 3 Sep 2011: 79-80. Accessed at http://www.economist.com/node/21528216 on 21 Sep 2011.

[229] Ryan, "*A Critical Analysis of Current and Proposed Future Nuclear Reactor Designs.*" See also T.B. Cochran, "Critique of 'The Future of Nuclear Power: An Interdisciplinary MIT Study.'" Accessed at http://www.pewclimate.org/docUploads/10-50_Cochran.pdf on 11 Oct 2011.

[230] Center for Climate and Energy Solutions, *Climate Techbook/Electricity/Hydropower.*

[231] "Total System Power for 2011," California Energy Commission, Aug 2012. Accessed at http://energyalmanac.ca.gov/electricity/total_system_power.html on 18 Nov 2012.

[232] Center for Climate and Energy Solutions, *Climate Techbook/Electricity/Biopower.*

[233] Ibid.

[234] "Biopower" in "Electricity from Renewable Resources: Status, Prospects, and Impediments," *Proceedings of the National Academy of Sciences (PNAS)* (2010): 104-11. Accessed at http://www.nap.edu/catalog.php?record_id=12619 on 29 Jun 2011.

[235] Center for Climate and Energy Solutions, *Climate Techbook/Electricity/Biopower.*

[236] Ibid.

[237] CleanTech Ostergotland, "Welcome to the Twin Cities of Sweden — Linköping and Norrköping," Accessed at http://www.cleantechostergotland.se/images/CTO_modellen.pdf. on 7 Jul 2011.

[238] C. Williams, "Power Plants," *New Scientist*, 11 Feb 2012: 46-49.

[239] "Geothermal Power" in "Electricity from Renewable Resources: Status, Prospects, and Impediments," *Proceedings of the National Academy of Sciences*

(2010): 92-7. Accessed at http://www.nap.edu/catalog.php?record_id=12619 on 29 Jun 2011.

[240] Jefferson Tester, et. al., *The Future of Geothermal Energy: Impact of Enhanced Geothermal Systems (EGS) on the United States in the 21st Century*, Massachusetts Institute of Technology (2006). Accessed at http://www1.eere.energy.gov/geothermal/pdfs/future_geo_energy.pdf.

[241] Energy Information Administration, *Electric Power Monthly*, 2011.

[242] USGS Newsroom, "Substantial Power Generation from Domestic Geothermal Resources," US Geological Survey, 29 Sep 2008. Accessed at http://www.usgs.gov/newsroom/article.asp?ID=2027 on 8 Apr 2012.

[243] Energy Information Administration, *Annual Energy Outlook 2011. Accessed at* http://www.eia.gov/forecasts/aeo/tables_ref.cfm on 30 Jun 2011.

[244] International Energy Agency (IEA), *Wind Energy Annual Report 2009*, p. 5. Accessed at http://www.ieawind.org/AnnualReports_PDF/2009/2009AR_92210.pdf .

[245] Platts.com, *Wind power installation slowed in 2010, outlook for 2011 stronger: AWEA*, 24 Jan 2011. Accessed at http://www.platts.com/RSSFeedDetailedNews/RSSFeed/ElectricPower/6773195 .

[246] IEA, *Wind Energy Annual Report 2009*, 6.

[247] Energy Information Administration, *State Renewable Electricity Profiles*, 8 Mar 2012. Accessed at http://www.eia.gov/renewable/state/ on 17 Aug 2012.

[248] Department of Energy (DOE), *20% Wind Energy by 2030: Increasing Wind Energy's Contribution to US Electricity Supply*, July 2008. Accessed at http://www1.eere.energy.gov/windandhydro/pdfs/41869.pdf.

[249] Ibid., 6 and 11, Fig. 1-9.

[250] IEA, *Wind Energy Annual Report 2009*, 158.

[251] R. Wiser, G. Barbose, and C. Peterman, *Tracking the Sun: The Installed Cost of Photovoltaics in the US from 1998-2007*, Lawrence Berkeley National Laboratory, Report No. LBNL-1516E, 2009. Accessed at http://eetd.lbl.gov/ea/ems/reports/lbnl-1516e.pdf .

[252] Center for Climate and Energy Solutions (C2ES), *Climate Techbook/Electricity/Solar Power*. Accessed at http://www.c2es.org/technology/factsheet/solar on 10 Mar 2012.

[253] D.R. Baker, "Renewable-Energy Bill Passed by Legislature," *San Francisco Chronicle*, 30 Mar 2011. Accessed at http://www.sfgate.com/cgi-bin/article.cgi?f=/c/a/2011/03/30/MN4L1IM38G.DTL .

[254] MIT News, "Where the Wind Blows," The Joint Program on the Science and Policy of Global Change, 25 Oct 2011. Accessed at http://globalchange.mit.edu/news/news-item.php?id=136 on 5 Nov 2011.

[255] Ibid.

[256] D.W. Kieth, "The Influence of Large-Scale Wind Power on Global Climate," *Proceedings of the National Academy of Sciences (PNAS)* 101, (16 Nov 2004): 16115-20. Accessed at http://www.pnas.org/content/101/46/16115.full.pdf+html?sid=55c39d29-5a92-424e-91b1-0da2f401b9ce .

[257] IEA, *Wind Energy Annual Report 2009*, 160-1.

[258] DOE, *20% Wind Energy by 2030*, 11.

[259] IEA, *Wind Energy Annual Report 2009*, 34.

[260] X. Lu, et al., "Global Potential for Wind-Generated Electricity," *Proceedings of the National Academy of Sciences (PNAS))* 106, no. 27 (Jul 2009): 10933-38, p10937. Accessed at http://www.pnas.org/content/106/27/10933.full.pdf+html .

[261] Ibid.

[262] R. Jackson, and C. Haddox, "Phantom Eye High Altitude Long Endurance Aircraft Unveiled," Boeing Inc. Accessed at http://www.boeing.com/Features/2010/07/bds_feat_phantom_eye_07_12_10.html .

[263] DOE, *20% Wind Energy by 2030*, 80.

[264] D. MacKay, *Sustainable Energy Without the Hot Air*, UIT Cambridge (2009): 186-202. Also available at http://www.inference.phy.cam.ac.uk/sustainable/book/tex/ps/113.252.pdf

[265] California Council on Science and Technology, "California's Energy Future: The View to 2050, Summary Report," May 2011: 26 and 34. Accessed at http://www.ccst.us/publications/2011/2011energy.pdf on 25 May 2011.

[266] Ibid., 194-5 and 198.

[267] Ibid., 158.

[268] US Department of Energy, "Grid 2030 - A National Vision for Electricity's Second 100 Years," *Office of Electricity Delivery & Energy Reliability*. Accessed

at http://energy.gov/oe/downloads/grid-2030-national-vision-electricity-s-second-100-years on 10 Jul 2012.

[269] US Department of Energy, "Department of Energy Announces Five Awards to Modernize the Nation's Electric Grid," *Energy.gov*. Accessed at http://energy.gov/articles/department-energy-announces-five-awards-modernize-nations-electric-grid on 10 Jul 2012.

[270] *"The Future of Energy," The Economist*, 19 Jun 2008, several articles. In particular see *"The Future of Energy, Trade Winds."* Accessed at http://www.economist.com/node/11565667?story_id=E1_TTVGVGGS.

[271] EIA, "Levelized Cost of New Generation Resources in the Annual Energy Outlook 2012," 25 Jun 2012. Accessed at http://www.eia.gov/forecasts/aeo/electricity_generation.cfm on 4 Oct 2012.

[272] EIA, *Electric Power Monthly*, Table ES1.B, March 2011. Accessed at http://www.eia.doe.gov/cneaf/electricity/epm/epm_sum.html.

[273] Center for Climate and Energy Solutions, *Climate Techbook/Electricity/Solar Power*. Accessed at http://www.c2es.org/technology/factsheet/solar on 10 Mar 2012.

[274] Solar power installations of over one megawatt account for 0.3% of California's total power generation (California Energy Commission, *Total Electricity System Power*, 2010). The other 0.7% comes from rooftop collectors, which recently topped one gigawatt. D.R. Baker, "State's Solar Energy Output Reaches Milestone," *San Francisco Chronicle*, 9 Nov 2011: D1. Accessed at http://www.sfgate.com/cgi-bin/article.cgi?f=/c/a/2011/11/09/BU6D1LS4E6.DTL on 9 Nov 2011.

[275] Sun Electronics at http://www.sunelec.com/ .

[276] Solar Power Panels, "Biggest Solar Panel Manufacturers in the World." Accessed at http://solarpowerpanels.ws/solar-panels/biggest-solar-panel-manufacturers-in-the-world on 6 *Apr* 2011.

[277] M. Buchanan, "Is Solar Electricity the Answer?" *New Scientist*, 2 Apr 2011: 8. Accessed at http://www.newscientist.com/article/mg21028063.300-wind-and-wave-farms-could-affect-earths-energy-balance.html .

[278] Ibid.

[279] T. Woody, "$4.1 Billion in Orders for Thin-Film Solar," *New York Times*, 9 Sep 2009. Accessed at http://green.blogs.nytimes.com/2009/09/09/41-billion-in-orders-for-thin-film-solar/ on 8 Apr 2011.

[280] G. Barbose, et al., *Tracking the Sun III: The Installed Cost of Photovoltaics in the US from 1998-2009*, Dec 2010. Accessed at http://eetd.lbl.gov/ea/ems/reports/lbnl-4121e.pdf on 8 Apr 2011.

[281] Rocky Mountain Institute, "Achieving Low-Cost Solar PV: Industry Workshop Recommendations for Near-Term Balance of System Cost Reductions." Accessed at http://www.rmi.org/Content/Files/BOSReport.pdf on 7 Jun 2011.

[282] Ibid. Figure 8.

[283] Google.org. "Plug into a Greener Grid: RE<C." Accessed at http://www.google.org/rec.html on 7 Apr 2011.

[284] Center for Climate and Energy Solutions (C2ES), *Climate Techbook/Electricity/Solar Power*.

[285] P. McKenna, "Coal is Far Deadlier than Nuclear Power," *New Scientist*, 26 Mar 2011.

[286] US Energy Information Administration (EIA), "Electricity: Non-OECD Asia," *International Energy Outlook 2011*. Accessed at http://www.eia.gov/forecasts/ieo/electricity.cfm on 23 Mar 2012.

[287] Ibid.

[288] EIA, "World Installed Solar Generating Capacity," Interactive table viewer on *International Energy Outlook*, 2011. Accessed at http://www.eia.gov/oiaf/aeo/tablebrowser/#release=IEO2011&subject=0-IEO2011&table=25-IEO2011&region=0-0&cases=Reference-0504a_1630 on 10 Jul 2012.

[289] EIA, *International Energy Outlook 2011*, "Electricity: OECD Europe."

[290] E. Shogren, "A Lease On Solar Panels? Some States Pitch In," National Public Radio, 22 Feb 2011. Accessed at http://www.npr.org/2011/02/22/133870498/a-lease-on-solar-panels-some-states-pitch-in .

[291] SolarCity at http://www.solarcity.com/ .

[292] Sungevity at http://www.sungevity.com/ .

[293] "Secrets of Residential Solar Lease—Sweet Deal or Disastrous Rip-off?" *San Jose Green Homes*. Accessed at http://sanjosegreenhome.com/2010/01/27/secrets-of-residential-solar-lease-sweet-deal-or-disastrous-rip-off/ on 30 Mar 2012.

[294] BrightSource at http://www.brightsourceenergy.com/projects/ivanpah .

[295] D.R. Baker, "US, Google Back BrightSource Mojave Solar Plants," *San Francisco Chronicle*, 12 Apr 2011. Accessed at http://www.sfgate.com/cgi-bin/article.cgi?f=/c/a/2011/04/12/BUAG1IUER0.DTL .

[296] R. Costa, "Thermal Storage Analysis in CSP Plants," *Global Solar Thermal Energy Council*, 20 Oct 2009. Accessed at http://www.solarthermalworld.org/node/874 .

[297] BrightSource website.

[298] Fred Pearce, "Current Power: New Tide Turbines Tap Oceans of Energy," *New Scientist*, 20 Sep 2011: 48-51. Accessed at http://www.newscientist.com/article/mg21128301.900-current-power-new-tide-turbines-tap-oceans-of-energy.html on 8 Oct 2011.

[299] MacKay, *Sustainable Energy Without the Hot Air*, 84.

[300] Center for Climate and Energy Solutions, *Climate Techbook/Electricity/Smart Grid*.

[301] Electric Power Research Institute, "The Green Grid: Energy Savings and Carbon Emissions Reductions Enabled by a Smart Grid," June 2008. Accessed at http://www.smartgridnews.com/artman/uploads/1/SGNR_2009_EPRI_Green_Grid_June_2008.pdf in Apr 2011.

[302] Center for Climate and Energy Solutions, *Climate Techbook/Electricity/Smart Grid*.

[303] US Department of Energy, *"Grid 2030 - A National Vision for Electricity's Second 100 Years," Office of Electricity Delivery & Energy Reliability*. Accessed at http://energy.gov/oe/downloads/grid-2030-national-vision-electricity-s-second-100-years on 10 Jul 2012.

[304] US Department of Energy, "Department of Energy Announces Five Awards to Modernize the Nation's Electric Grid," *Energy.gov*. Accessed at http://energy.gov/articles/department-energy-announces-five-awards-modernize-nations-electric-grid on 10 Jul 2012.

[305] D. Talbot, "Lifeline for Renewable Power," *Technology Review*, Jan/Feb 2009, published by MIT. Accessed at http://www.technologyreview.com/energy/21747 on 21 Oct 2011.

About the author

Robert Fraser is a consulting scientist at Foresight Science & Technology. His research over the years has focused on the development of instrumentation with applications in medicine and environmental monitoring. He received his undergraduate degree in Physics from the University of Notre Dame and his doctorate in Fluid Dynamics from the University of California at Berkeley.

Made in the USA
Charleston, SC
25 March 2013